T0340102

Energy Efficiency in Air Transportation

Energy Efficiency in Air Transportation

Arturo Benito
Gustavo Alonso

Butterworth-Heinemann is an imprint of Elsevier
The Boulevard, Langford Lane, Kidlington, Oxford OX5 1GB, United Kingdom
50 Hampshire Street, 5th Floor, Cambridge, MA 02139, United States

Library of Congress Cataloging-in-Publication Data
A catalog record for this book is available from the Library of Congress

British Library Cataloguing-in-Publication Data
A catalogue record for this book is available from the British Library

ISBN: 978-0-12-812581-6

For information on all Butterworth-Heinemann publications
visit our website at https://www.elsevier.com/books-and-journals

Publisher: Matthew Deans
Acquisition Editor: Carrie Bolger
Editorial Project Manager: Carrie Bolger
Production Project Manager: R.Vijay Bharath
Cover Designer: Christian J Bilbow

Typeset by SPi Global, India

Contents

Introduction

1

Air transport is without any discussion one of the most relevant components of the social organization of our world. In 2017, there were about 4200 million of commercial air trips, more than one per every two-world inhabitants and the majority of the air traffic forecasts predict growth trends in the order of 5%–6% for the next 20 years, clearly higher than the expected economy increase rates.

The air cargo market is also important in a somewhat different way. In terms of weight, the demand is much smaller than in other transportation means, not reaching 0.1% of the ton-kilometer transported, but in terms of the value of the goods, it becomes close to 30% of all the import-export world market. The possibility of moving high-value components in a fast and safe way has modified the producers' behavior and the consumers' as well.

In economic terms, world airlines revenue went up to 754 billion USD, a figure that would put commercial aviation as number 19 of the World States Gross National Product (GNP) ranking, comparable with the Netherlands or Switzerland. However, the global effect is much greater, supporting 3.5% of global Gross Domestic Product (GDP), equivalent to 3.3 trillion USD, and summing up to 69.6 million jobs, if direct, indirect, induced, and catalytic effects are computed. This activity generated 126 billion USD in different taxes going to the national economies.

A very relevant factor is the increasing connectivity provided by air transport. In 2017, world airlines flew more than 20,000 unique city pairs, a figure more than double with respect to the year 1996. With a route structure showing a high level of interconnection in the airports acting as hubs and a growing number of intermodal relationship, modern air transport has a unique capability to put in contact the world population, serving its social and economic demands.

This upwards trend is confirmed by a number of other economic and behavioral evidences of our present society, as the widening of globalization, the growth of tourism, and the increasing interest for traveling of the new generations. All those phenomena needing of air transport to develop their maximum potentialities and requiring a continuous air transport offer enlargement in order to satisfy the existing and new developing demand. The strong and accelerated economic growth of some of the most populated countries, like China, India, Indonesia, or Brazil, is asking for the support of a well-balanced air transport network, able to support the domestic development and link it with the surrounding economies worldwide.

In comparison with the majority of activities in the services sector, including other transportation modes, commercial aviation is energy intensive and this feature makes mandatory the introduction of every mean to minimize its energy use due to economic, environmental, and sustainability reasons. The relative importance of these three elements has changed along the years, following the fluctuations of the world economy, the sociopolitical developments of different world regions, and the discovering

Energy Efficiency in Air Transportation. https://doi.org/10.1016/B978-0-12-812581-6.00001-6

of new scientific evidences on the evolution of the Earth climate. In any case, the global weight of the three factors has continued going up and putting more pressure on the different sector stakeholders: on manufacturers for developing more efficient technologies, to be adequately tested and brought into commercial service conditions; on operators for making the best possible use of those new vehicles in the general operative context; and on regulators for introducing new measures to optimize energy use and making a more sustainable operative landscape.

In this specific transportation mode, the problem is more acute due to the absence of short/medium term alternatives to the use of oil products (kerosene and high octane gasolines) as standard fuels. Longer-term options, like liquid hydrogen, would require a drastic change in the system structure and look very far away in time. While ground and sea transport may access gas, electric, solar, wind, or even nuclear energy, the air mode is just starting to analyze batteries, solar panels, and fuel cells as possible auxiliary energy providers and the only medium-term technically feasible alternative source appears to be the kerosene obtained from biological feedstock. However, its present price in today's market conditions makes this option unfeasible, unless regulatory actions are taken to introduce these substances in the industry through mandatory or economic incentive measures.

All aforementioned circumstances taken into account, it seems pertinent to gather in a book those many different aspects of the energy efficiency in air transport. Not only to provide answers to questions coming from diverse areas of the sector but also, perhaps, to propose some new research items that may appear in the next future, as results of the advanced technology now in preliminary status. This book intends to provide the necessary background information, together with the newest developments in a very dynamic field of knowledge, and become a useful tool for air transport professionals, decision makers, and university students wishing to specialize in this area or to find interesting data to be used in other different but related research tasks.

The structure of the book starts with this introductory section (named as Chapter 1), and follows with other nine chapters, covering the energy efficiency aspects of the three main elements of the air transport system (aircraft, air space, and airports) design and operation, and their respective impacts on environment and sustainability. The volume closes with a very wide and exhaustive list of references that have been used along the different chapter expositions.

The second chapter intends to describe the framework in which the air transport energy consumption is inserted. It starts with a wide panorama of the world present energetic situation, showing the primary energy production sources and the main consumption elements, with a reference to the most likely future developments. An overwhelming majority of the fuel used in air transport is coming from crude oil distillation, taking the form of kerosene, but there are other minor fuel sources, like organic feedstock, waste, or carbon that may be important in the future. This chapter describes, in technical and economic terms, the existing and potential sources and the similitudes and differences with other transportation means, indicating the world oil consumption share of the air transport and the comparative efficiency in competitive or uncompetitive trips.

Chapter 3 describes the core structure of the air transport system, detailing its main components and their respective importance for the optimization of energy consumption. The air transport mode is defined by three basic elements: commercial transport aircraft, the air space used for their flights, and the infrastructure required to supporting their operations, like airports and air traffic management (ATM) facilities. The large majority of energy consumption is coming from the aircraft operation and the chapter covers the composition of the 28,000-strong commercial aircraft fleet, with a brief description of existing aircraft and engine types and the prospective future programs. The other two stakeholders have a substantial influence in the global result as well. Airports, in particular, have a different energy balance and offer more chances to use alternative energy sources and achieve a zero-carbon footprint, but the way in which they manage their traffic has a large effect in the fuel consumption by the operators. In the case of ATM, it is obvious its decisive influence on optimizing the trajectory and flight regime of the operations in order to optimize the energy bill. Over these three elements, the regulatory system has an obvious relevancy because a great number of operational regulations have direct effects on the energy use and the operational efficiency.

The design of commercial aircraft and the effects of different elements design in fuel efficiency are the issues dealt with in Chapter 4, divided into four subchapters: aerodynamics, propulsion, structures, and systems, corresponding to a typical subdivision of the aircraft architecture. In each one of them, the main existing and proposed features to reduce energy consumption are detailed, including in service and in development applications, and medium-long term expectations, like nonconventional aircraft or engine configurations. A second part of this chapter provides some information on the evolution of the energy efficiency in the last years, in relation with the aircraft-type replacement. A final section reviews some of the most important technological programs going ahead in Europe and United States.

Chapter 5 moves to the daily operation of commercial aircraft and the fuel efficiency consequences of the preparation of commercial service flight planning, with the relative importance of different parameters, like flight speed, aircraft weight, flight track, or cruise altitude. The repercussions of last-minute unexpected events on the flight plan and the different possibilities of managing those circumstances are also discussed. An analysis of most commonly used optimization policies at the moment of elaborating the flight plan, including the application of Cost Index, tankering, redispatch/reclearance, and flight management system (FMS), helps to go from the theory to the airline practice of each individual flight dispatch.

Next chapter presents the detailed fuel efficiency effects of each flight phase (ground operations, take-off, climb, cruise, descent, approach, landing, and taxi) and the optimization procedures, including the elements depending on the airport procedures and air navigation services operation. Trajectory optimization and the application of the ATM instructions are also contemplated and discussed.

The importance of sound maintenance practices to optimize fuel use is analyzed in Chapter 7. The two main issues to consider are the systems for aircraft performance monitoring and the maintenance actions focused on improving fuel efficiency.

Aircraft and engines performance monitoring is a valuable decision tool to follow the performance degradation and to evaluate the different possibilities of performance/reliability recovery actions to be added to the standard maintenance program. To put those actions into value, their potential fuel efficiency effects need to be compared, by the adequate tradeoffs with the changes in other operational conditions, like aircraft availability for commercial service, on-time performance, and global flight economy. This chapter makes also a smart tour around the maintenance actions with the most important effects on fuel efficiency, like engine refurbishment, aerodynamic cleanness, and weight reduction.

Chapter 8 pays a fast visit to the energy footprint of the air transport infrastructure. Although the share of energy consumption of airports and ATM facilities is very small, compared with the aircraft movements, and they have more possibilities of using alternative energy sources, their correct operation may have important repercussions in the flight optimization. The effects of infrastructure in energy use are divided into two parts: the first one deals with ATM developments to optimize flight tracks and reduce congestion, including the restructuration of the airspace and the international cooperation on flight tracking analysis under International Civil Aviation Organization (ICAO) surveillance. The second explores the use of energy in the airport facilities and the effects of airport operations both to the airlines and to the airport itself. This part reviews the airside of the airport, the ground one and the other transportation modes access to the facilities, with a final note on the airport carbon certification system.

The relationship between aviation energy efficiency and the environment protection is described in Chapter 9. The aviation environmental impact from energy use has two different parts: local effects (air quality, energy management facilities in the airports, landscape alteration) and global effects (consumption of nonrenewable materials, use of dangerous substances and, the most important, contribution to climate change). In addition to the technical actions exposed in the previous chapters, an analysis of the potential of market-based measures (taxes, charges, voluntary agreements, emissions offsetting, and emissions trading systems) to reduce this impact is performed. Due to the complicated regulatory framework of the aviation environment, a list of the most important regulatory bodies and the key laws and regulations is included, with a mention to the Environmental Management Systems and their energetic repercussions.

Chapter 10 deals with the most recent developments in the area of the efficiency certification. While a number of commercial aircraft features require a certification under the design state rules before entering into service, fuel efficiency had not been one of the certification parameters until now. However, after complicated and hectic six-year-long discussions, the 2016 ICAO General Assembly approved a CO_2 emissions certification requirement for new aircraft models, achieving type certification in 2020 and onwards, with a later extension to models in production. During the same meeting, ICAO also approved the implementation of the Carbon Offsetting and Reduction Scheme for International Aviation (CORSIA), intended to keep net international civil aviation CO_2 emissions at a level not higher than the one produced in the year 2020. The technical bases of the certification system are described here as well as the general principles of the CORSIA system, but many detailed features are still in discussion within the ICAO Committee on Aviation Environmental Protection (CAEP).

Energy and air transport 2

2.1 Energy and transportation

Transportation means moving massive things from one point to other and requires a certain amount of energy. The dimension of the energetic needs depends on the achieved momentum (mass and speed), the vehicle technology, the medium in which transport is done, the infrastructure qualities, and the legal regulation applicable.

Any transport consumes energy in the quantity demanded by those previously mentioned elements. This energy has a number of conditions to adapt itself to the type of movement of each different transportation mode. The final fuel product has to arrive to the engine in the most convenient condition and, at the same time, production and logistics should not be excessively complicated, expensive, or demanding a large amount of additional energy. In statistical terms, the majority of transportation energy figures is used to refer to the final consumption and may not give proper attention to the total energy spent in the global process. The difference between assessing final consumption and calculating a complete life-cycle analysis may be very big and change totally the desired conclusions.

In some cases, the fuel used to extract the energy is coming directly from a natural source, like the mineral oil distillated into gasoline or kerosene. In others, there is an intermediate conversion before the final consumption, like the train using electricity produced by a solar, wind, or hydroelectric plant, or a mix of all of them. Many of the transportation vehicles, like cars, ships, or planes, are carrying the fuel inside, while others, like electric trains, take energy from a continuous line. This makes a lot of difference because taking the fuel on board means that the vehicle-operating weight changes during the trip and the efficiency is not constant along the same travel. At the same time, continuous energy feeding is producing additional losses with respect to the basic fuel logistic distribution.

Our present type of social development has been closely tied to the use of growing amounts of energy in the last couple of centuries. In the early times of the civilization, the energy sources were from natural origin to make fire (wood, dry plants, coal, vegetal oil), leaving transport activities to the power of the wind (sails) or the muscular energy from men (walking or rowing) or animals (horses, oxen, camels, dogs, or even elephants). The great jump ahead in energy generation, transport and final use happened in the 19 century, with the introduction of vapor machines and internal combustion engines, both burning mainly fossil fuels. There were immediate repercussions for the industrial development and the expansion of the transportation networks, pushing up the energetic demand to produce goods and distribute them in the world market. An additional spin-off was the search of additional fossil fuels in addition to the already-extended coal.

The generalization of the electricity use along the 20 century put more pressure on the energy demand for fixed installations, in parallel with an exponential increase in

Energy Efficiency in Air Transportation. https://doi.org/10.1016/B978-0-12-812581-6.00002-8

the number of private road vehicles, bringing transportation up to a 25% of the world energy consumption. The dependence on oil production became headline news after the first oil price sharp increase, a consequence of the Arab-Israeli Yom Kippur war in 1973. The high level of fossil-fuels consumption provoked some warnings on the economic effects of expensive energy and the probable scarcity of world oil and gas reserves. Some years after, a new factor joined those two: the contribution of use of energy-related emissions to the climate warming might make impossible the stabilization of the atmospheric temperature at an average level not more than 2°C higher than in preindustrial times, a level required by the United Nations Framework Convention on Climate Change (UNFCCC) in order to avoid severe consequences on our climatic welfare.

Great research efforts have been made to diversify energy sources and find other products capable to replace, at least partially, the oil dependence, something very important in many human activity areas but particularly serious for transportation. The capability of producing energy at competitive conditions would need to be complemented by the feature of being renewable and not generate a high level of greenhouse gases. Table 2.1 shows the existing inventory of energy sources, being used or researched, within the limits of our present technology. They are classified into two groups, according to the renewability feature and, in each case, it is indicated the source from which it is extracted, the class of production, the resulting energetic substance, and the type of processing needed in order to obtain the energy.

The quality of being renewable means that the basic material producing energy can be replaced in a reasonable time scale. Oil and natural gas are products of millions of year long process. According to different sources, at the end of 2016, the proved oil reserves were equal to 50.6 years of production at that year levels. In the case of natural gas, the figure of proved reserves was very similar: 52.5 years, with the difference

Table 2.1 Types of energy

	Source	Class	Substance	Energy process
Nonrenewable	Mineral	Fossil	Oil, gas, coal	Combustion
		Nuclear	Uranium	Fission reaction
			Hydrogen	Fussion reaction
		Geologic	Internal heat	Conversion
	Solar	Active	Radiation	Photovoltaic
				Thermal
		Passive	Radiation	Bioclimatic architecture
	Air	Eolic	Wind	Mechanic
Renewable	Gravity	Hydraulic	Water	Mechanic
		Tidal, Waves	Water	Mechanic
	Organic	Vegetal	Biomass	Combustion
		Used elements	Waste	Combustion
	Biological	Life energy	Muscle	Oxidization

that oil demand is more or less stabilized while the amount of gas delivered continues increasing at fast pace. Proven reserves volume is not price independent. As oil barrel becomes more expensive, some deposits or extraction procedures, previously considered as not economically viable, become acceptable and the reserve level increases. High sea oil extraction or fracking are good examples of expensive extraction procedures only profitable at high oil barrel prices.

The other nonrenewable source is the nuclear energy, either the already-industrialized fission technique or the researched fusion procedure. The highly publicized safety concerns and the difficulties for radioactive waste treatment and storage make unlikely to think of a fast increase of nuclear energy in the near future. Some States, like Germany or Switzerland, have committed on the closing of their nuclear power plants, and some others have moratorium on the building of new ones. Finally, the modest scale of geothermic energy use seems to pose no exhausting problem in this area, but its peculiarities for treatment and control limit the range of potential applications.

Renewable energies are more heterogeneous and the majority of them still move in the first part of the learning curve. Solar radiation is clearly the most promising as the solar cells efficiency factor increases with experience and research. Mechanical technologies use the power of natural elements, like water, wind, tides, or waves, to convert them into electricity. The oldest of those technologies is the hydraulic production of electricity, building dams in the parts of the rivers with an adequate orography, having each about 8% of the total world energy consumption. It is a very high figure but not comparable with oil, natural gas, and coal that dominate the global markets, as can be seen in Fig. 2.1, showing the evolution of the world energy production in the last 25 years. Vertical scale is in Toe (Tons of oil equivalent) and hydroelectric generation has been split from the rest of renewable sources, with the purpose of segregating a source that has been utilized since long time ago from the other renewable technologies, being relatively newcomers.

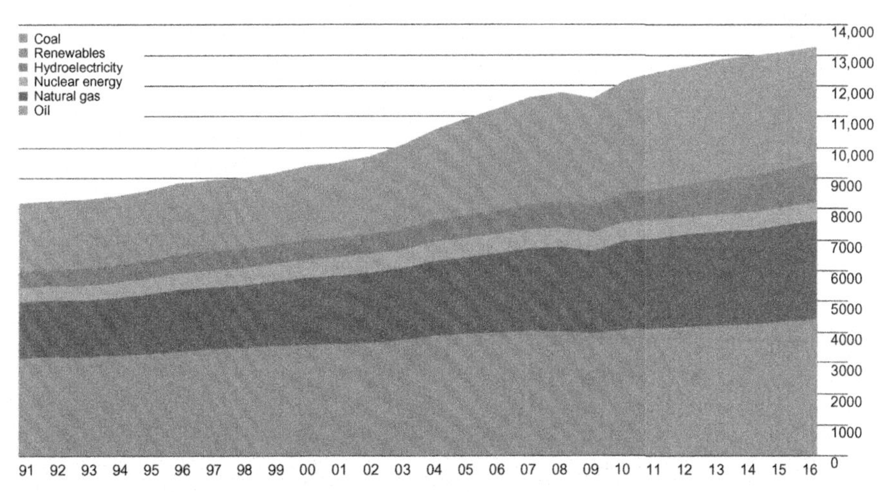

Fig. 2.1 World energy consumption evolution in Toe.
Source: BP Statistical Review of World Energy

The effect of some of the renewable energies is fairly stable, as it is the case of solar radiation, sea waves, or tides, but others are dependent on climatic conditions. The amount of rain in a certain season directly affects the water volume in the dams. Too little water means low energy production, while too much requires to open floodgates without taking advantage of all its potential energy. A similar reasoning is applicable to wind energy, the supply of which may change depending on weather conditions. The periodical variations of those productions shows the problem of how to store them, a technique that involves the conversion to a different kind of energy, like electricity or heat.

In the last 50 years, there have been some changes in the distribution of the energy market, the evolution of which is indicated in Fig. 2.2, in percentages of the total. The share of oil and coal decreased from 42% to 36%, respectively, in 1966, down to 33% and 28% in 2016. This reduction in nonrenewable energy was partially compensated by natural gas increasing from 16% in 1966 up to 23% in 2016, and the beginning of the nuclear power, taking off in the early 70's and reaching about 5% last year. Hydroelectric was stable around 6-8% and the rest of the renewable energies, which were below 1% until 10 years ago, is now at 3.2% of global primary energy consumption.

An accurate evaluation of the primary energy consumption distribution among the most important economic sectors is difficult because the definition of those sectors is not homogeneous in different parts of the world. The International Energy Agency (IEA, Fig. 2.3) gives to transportation a global 35% for 2014, passenger cars being 60% of that number. The U.S. Energy Information Administration offers the 29% figure for the year 2016 in United States, while Eurostat says 33.1% in 2015 for the 28 members of the European Union. All the references agree on considering that transportation share is increasing, as other sectors are faster and more economically efficient in adopting measures to reduce their energy consumption.

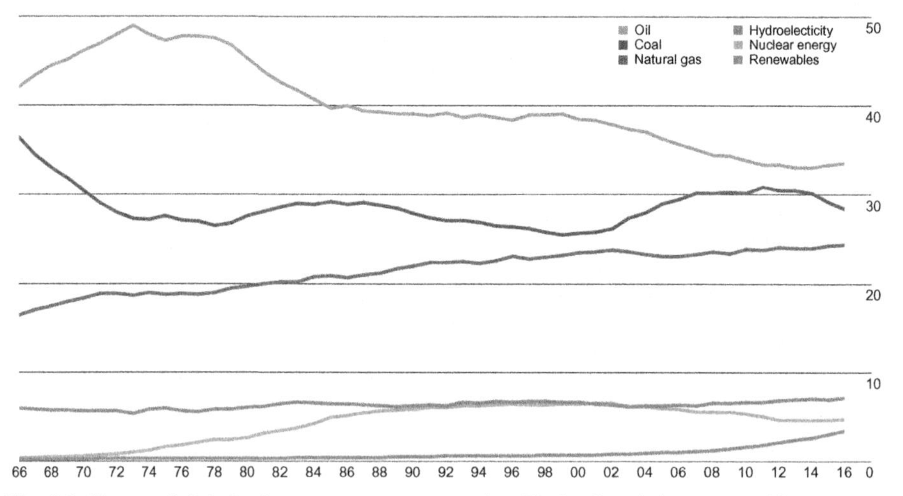

Fig. 2.2 Shares of global primary energy consumption. Hydroelectric is separated from the rest of renewables.
Source: BP Statistical Review of World Energy

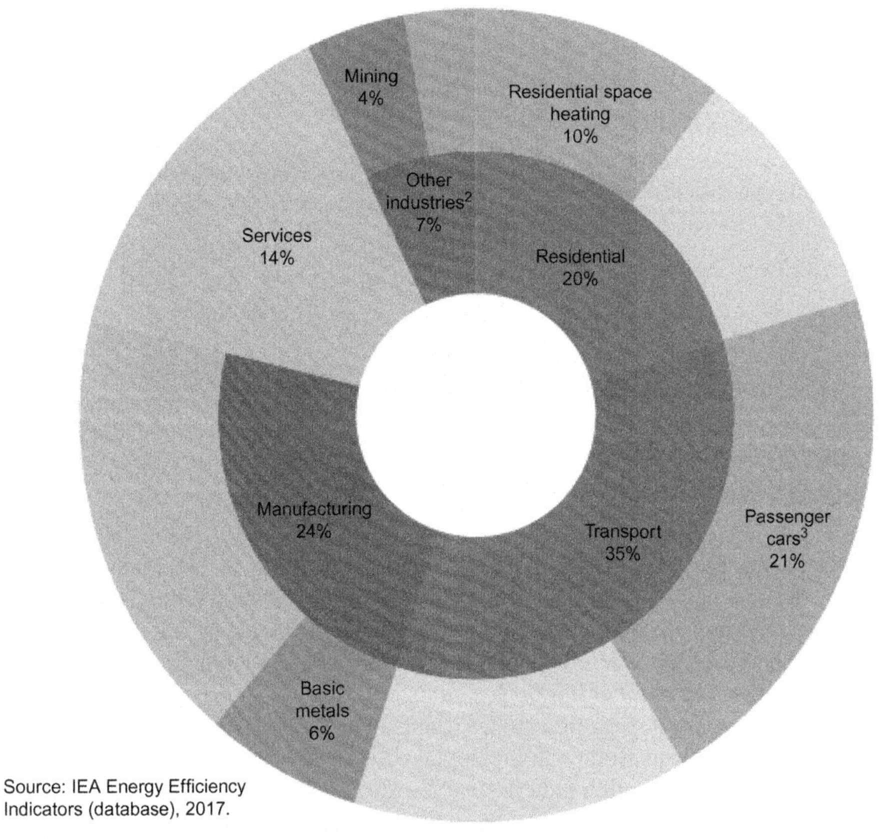

Fig. 2.3 World energy consumption by end-use sector (2014)

Transportation services compete for energy with other human activities, industry, agriculture, housing, etc. in the global market. As population grows and social development increases, the demand for energy continues going up in practically every region of the world, and the energy industry tries continuously to satisfy that demand using all the different sources available at competitive prices and with high levels of safety in their use.

Today, more than 90% of the energy used by transport modes is coming from the oil in form of gasoline, diesel, kerosene, or LPG (liquefied petroleum gas), with a small participation of electricity. Some of the projections to the next 25 years (Fig. 2.4) are conservative about the consolidation of that distribution. However, there is a trend among the most developed States to incentivize the use of electricity in the road transport;, with the purpose of alleviating the local air-quality problems in the big cities. Then, the electricity share in transportation energy may soon become greater than the one predicted in present forecast.

In this graph, biofuels are included in the gasoline and diesel fuels. The production of biofuels for mixing with road vehicles standard fuels is becoming sizeable, stimulated by some State incentives and regulatory actions prescribing a minimum mandatory biofuel content in the product sold to the public. In United States, during the year 2016, 14.4 billion gallons of fuel ethanol were added to motor gasoline

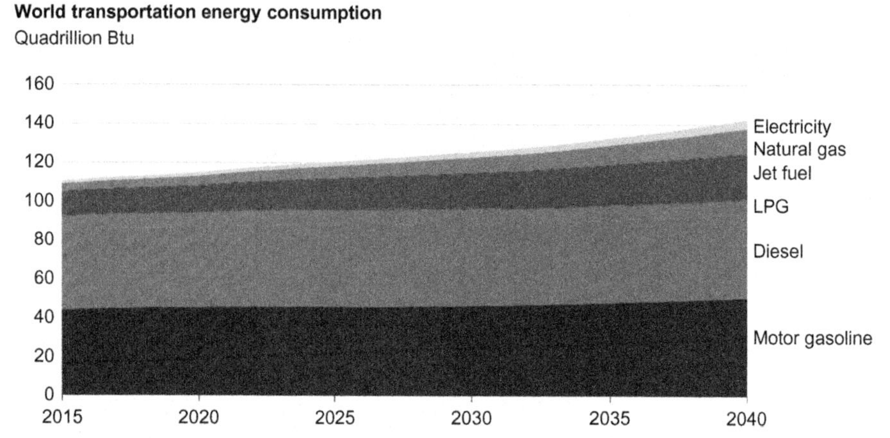

Fig. 2.4 Transport energy consumption projections until 2040.
Source: International Energy Outlook, U.S. Energy Information Administration.

consumed for transportation. That quantity is about 9.5% of the total sold fuel. Biodiesel consumption added another 2.1 billion gallon to the road transport figure. A more detailed explanation on regulatory procedures adopted by different States to incentive the use of biofuels, both in ground transportation and, potentially in aviation, can be found in Chapter 9 of this book.

The majority of transportation modes can use more than one source of energy, because their power plants work with a relatively modest power-to-weight ratio and there is no great volume limitation either. Road transport is the largest user of oil-derived fuels, either gasoline or diesel fuels, but it can accept gas, liquid gas, electricity, or a mix of them. Typical hybrid vehicles are gasoline/electrical, but other combinations, like gas/electric, are equally possible.

Trains are large electricity consumers, but diesel or coal are also suitable for medium and low speeds, while high speed trains (HST), between 250 and 350 km/h, are using exclusively high-voltage electricity (3000 V). Navigation is very variable, burning residual fuel oil, diesel, coal, liquid gas, or even nuclear energy or the classic windsailing. Natural gas is also used to operate compressors to move liquids in pipelines.

Unlike them, air transport is totally dependent on oil, in the shape of high-octane gasoline for piston engines (very small global consumption levels) and kerosene for turbine machines (jets, turboprops, and turboshafts) that make the great majority of the energy provision. The very strict technical requirements for aviation fuels make it difficult to find alternative products with comparable physical and chemical features, providing a good power-to-weight and power-to-volume relationships. Using 2016 data (Table 2.2), aviation kerosene and aviation gasoline take about 9% of the world oil of the year 2016 production, with 70% going to gasoline and diesel fuels.

All these oil-derived products go to different activities, but the part dedicated to the transportation has been increasing steadily in the last 50 years. Fig. 2.5 gives an idea of the distribution of the oil among different sectors, like transportation, industry, nonenergy use and other sectors and Fig. 2.6 shows the evolution of the shares of

Table 2.2 Petroleum products made from a barrel of crude oil (aprox. 45 US Gallons after refinery processing)

Oil destillation	Product	Use
4%	Liquefied petroleum gas	Heating, cooking, vehicles
45%	Automotive gasoline	Road transport
9%	Kerosene/aviation gasoline	aviation
25%	Diesel and heavy fuel	Road, trains, shipping
4%	Heating and heavy fuel oil	Housing/Industrial
13%	Other heavy fuel	Navigation, industrial, construction, diverse

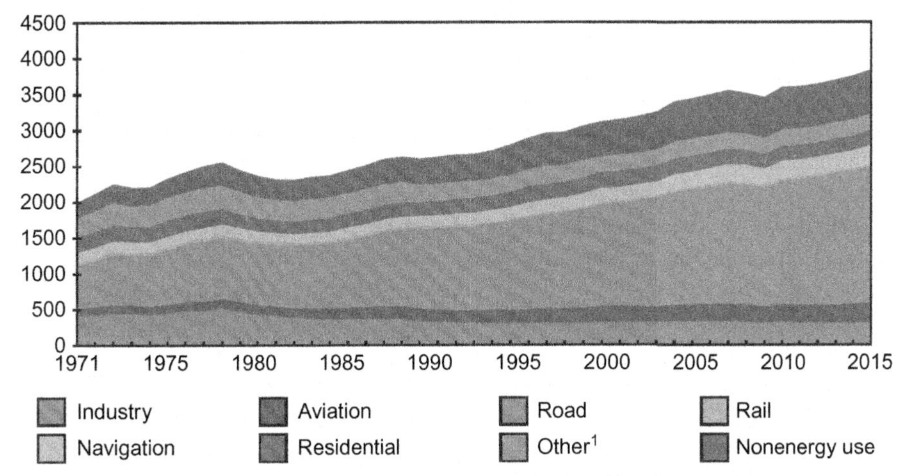

Fig. 2.5 Oil total final consumption by sector in MToe in the 1971–2015 period.
Source: International Energy Agency. Key world statistics 2017.

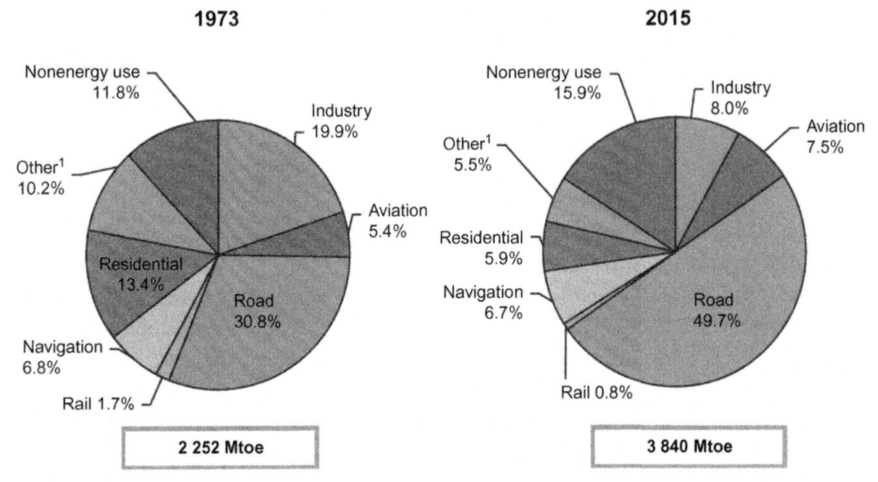

Fig. 2.6 Evolution of the different activity share in the total final oil consumption in the 1973–2015 period.
Source: International Energy Agency. Key world statistics 2017.

each activity sector in the 1973–2015 period. In the graph, nonenergy use refers to the manufacturing of products based in oil, like plastics or fertilizers. The sector named as others covers residential, commercial, public services, agriculture and a variety of smaller activities.

In these three decades, the road share has grown up 19 points, being now almost 60%, and aviation increases its share 2.1 points, enlarging almost 40%. On the other direction, industry, residential and diverse activities reduce their participation as oil derivatives are being replaced by electricity and natural gas.

2.2 Energetic efficiency of the transportation modes

The comparison of the energetic efficiency of different transportation modes is a difficult matter and subject of many scientific controversies. A number of issues are unclear and the way in which they are interpreted may lead to dissimilar and not comparable conclusions.

An initial issue is to determine whether the calculation target of the consumed energy should be the final energy used by the vehicle engine or should include the part of the cycle corresponding to the energy extraction and processing, and its transportation and delivery at the vehicle. This is very relevant for electricity that can be originated in many different ways from several sources, and is prone to suffer high levels of losses during transport and logistics distribution.

A second problem is the selection of the efficiency parameter that may be the number of transported passengers and/or weight of freight along a certain distance. The transportation subject is not problematic when it is relatively homogeneous, for example, number of individuals travelling but gets more debatable if baggage or freight is added and a mass number has to be assigned to the passengers in order of being added to the other transported materials. Different transportation modes use to have different standard weight(s) for passengers making comparisons difficult.

Other problematic issue is the trip distance, always conditioned by the available infrastructure, in such a way that travelling between two points may have very different trajectories depending on the transportation mode used. In Europe, for example, road distances are 20-25% longer than air distances and the calculated Revenue Passenger-Kilometer (RPK) or Revenue Tonne-Kilometer (RTK) for the same trip would not be the same. For conventional railways, deviations may be even greater, while HST has more direct tracks. If the selected unit for analysis is passenger-kilometer, correction factors are needed for a sound comparison.

The third question to consider is the scope of the accounted energy. Most calculations compute the final consumption at the vehicle, with or without the fuel life cycle, but in others, the energy consumed by the infrastructure (rails, stations, airports, highway services) is added to the final figure. This part of the energetic bill uses to be small in comparison with quantities corresponding to the vehicle, but the balance changes when the energy spent for the construction of the facilities is included. Ground transport infrastructure requires a lot of energy in its development and the global accounting is complicated but offers very different results than those coming from the simple final energy computation. In both cases, vehicle and infrastructures, there are some additional energy consumed in maintenance and operations activities.

Fig. 2.7 shows an energy efficiency analysis, made by Chester and Horvath, of three commercial aircraft models, using the life-cycle analysis methodology. Selected aircraft were representative of three different sizes and ranges: Embraer 145, 50 seater regional aircraft, Boeing 737, 150 seater medium-range aircraft, and Boeing 747, 400 seater long-range aircraft. The efficiency parameter is MJ/PMT (Mega Joules per Passenger-Mile Transported).

The energy assessment covers six areas:

- Aircraft operation, including different flight phases, like cruise, taxi, takeoff, climb out, approach, engines startup and auxiliary power unit (APU).
- Aircraft and engine manufacturing
- Insurance (calculated through Economic Input-Output Life-Cycle Analysis)
- Airport construction and operation
- Aircraft and airport maintenance
- Kerosene refining and distribution

The analysis results indicate that aircraft operation is between 80% and 85% of the total energy consumption, with other 7% going to the fuel processing and a small quantity to the aircraft and engine manufacturing. The energy consumption by the infrastructure related parts is practically negligible, because it is diluted in a large

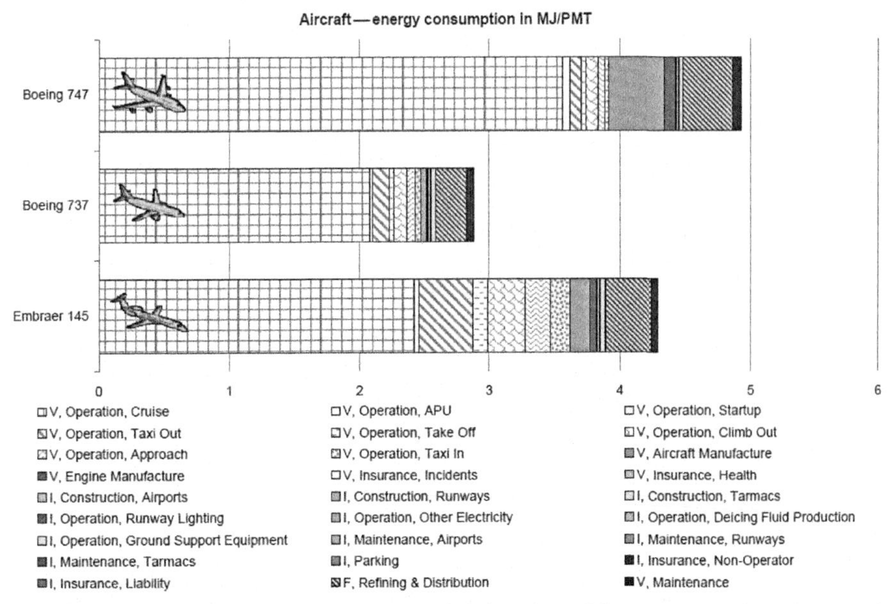

Fig. 2.7 Energetic efficiency of three commercial aircraft models under the Life cycle analysis approach.
Source: M. Chester and A. Horvath, UC Berkeley Center for Future Urban Transport, 2008.

volume of transportation production. This makes a great difference with road or railway modes in which construction, operation and maintenance of their respective infrastructures is significant in terms of energy consumption.

Table 2.3 presents the ranking of the domestic passenger transport in the United States. Energy efficiency is calculated as Mega Joules per RPK, using an average load factor. The most efficient mode is the Intercity train, in this case the Amtrak, taken as the reference and 8% better than the airplane, which is practically tied with the standard train. The suburban trains are 24% worse than the reference. After them, the road vehicles: motorcycles, cars utility cars and buses.

Two elements are the most important factors for arriving to these figures. First is the technology levels: Amtrak is clearly less efficient than the modern HST in Europe or Far East. Typical efficiency for the French TGV or the Spanish AVE are about 1.15 MJ/RPK, 20% better than Amtrak. On the other hand, airline efficiency in the US domestic traffic appears to be much better than in other zones of the world. The second key point is the vehicle load factor. The extremely low number of passengers per trip put the bus at the bottom of the classification in this analysis.

As an element of contrast, Fig. 2.8 shows IEA global statistics around the world, with the range of efficiencies of rail, bus, two-wheeled vehicles, cars and other personal vehicles, and air transport. In this case, the bus is the most efficient mode and the car and the airplane the highest energy consumers. As previously indicated, there are big differences in trains, with the intercity services being the best and the suburban services the worst. The differentiation in car consumption comes also from the great variety in vehicle types and, overall, from the load factors. Developed countries like USA or Europe have an average occupation between 1.2 and 1.4 passengers per trip, while in other parts of the world cars use to go with many more people inside.

In the same figure, there is a range of energy efficiencies in freight transportation, using Mega Joules per RTK. Shipping and rail advantage is considerable, with road in a distant third place. Air freight consumption is very high, but in the real life things are a little bit different because the standard calculations are made with cargo aircraft only and a majority of the air freight goes in the bellies of passenger airplanes. This is an

Table 2.3 **Energy use in the domestic U.S. transport in 2014**

Transport mode	Average passengers per vehicle	MJ per passenger-kilometer	Modal comparison
Intercity rail	20.9	1.433	Base
Air	108.9	1.553	+8.4%
Transit rail	26.7	1.561	+8.9%
Motorcycles	1.2	1.622	+13.2%
Commuter rail	31.6	1.775	+23.9%
Car	1.6	2.046	+42.8%
Personal trucks	1.8	2.335	+62.9%
Bus	9.2	2.510	+75.2%

Source: U.S. Transportation Energy Data Book Ed.35, July 2015.

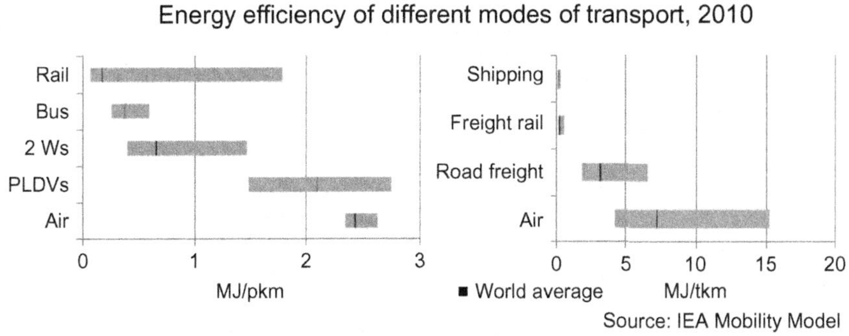

Fig. 2.8 Range of energy efficiencies of the different transportation modes in passenger and freight transportation.
Source: International Energy Agency, 2014.

aspect not solved in the intermodal comparisons. In passenger wide-body aircraft, belly cargo may take a sizeable part of the energy use and it is not taken into account in many calculations. There are some methods to include this effect, like to assign to belly cargo the part proportional to the consumption of the equivalent freighter, meaning the fuel needs of the same type of aircraft but in whole cargo configuration. These approaches add complication to the calculation and are not very popular in the intermodal analysis.

2.3 Air transport fuel economy

During the year 2017, world airlines consumed 88 billion US Gallons (266.5 M tons) of kerosene, at an approximated cost of 130 billion USD, what was 18.8% of their total expenses. The average price of a Brent oil barrel was 54.2 USD, equivalent to 63.3 USD per a similar size kerosene barrel.

The consumption figure is the maximum in the aviation history, but the others have been overpassed in previous years. The top of the economic cost was, up to now, the year 2013, when the total fuel bill was 230 billion USD (77% more than in 2017). The reason for that difference is the price of the oil during 2017, at levels corresponding to the year 2005 when marked an average price of 54.5 USD/barrel, with the additional factor that the exchange rate of the American currency with respect to the rest of the world was much higher at that time. The share of the kerosene in the total annual expenses of the airlines was 18.8%, again the minimum since the 16.0% in 2004, with an oil price of 38.3 USD per barrel. Table 2.4 reproduces the evolution of these four parameter during the last 12 years.

In the same table, the annual revenues and total operating profits are included. The price of fuel seems to have little effect on the volume of consumption, because demand is absorbing, with relative easiness, any related price increases. The ticket fare incorporates it either in the basic pricing or through specific surcharges with a mention to the fuel cost.

However, when the oil price gets down, there is an almost immediate decrease in revenues. A highly competitive market, as the air transport is, transfer speedily the

Table 2.4 Fuel consumption and expenses of the world airlines, compared with revenues and operating profits

	2005	2006	2007	2008	2009	2010	2011	2012	2013	2014	2015	2016	2017
Fuel bill (BUSD)	91	127	146	203	134	151	191	228	230	224	174	132	130
% total expenses	22.2	28.4	29.8	35.6	28.3	28.3	30.7	33.2	33.2	30.9	26.4	20.6	18.8
Brent USD/barrel	54.5	65.1	73.0	99.0	62.0	79.4	111.2	111.8	108.8	99.9	53.9	44.6	54.2
Consumption (BUSG)	68	69	71	70	66	70	72	73	74	77	81	85	88
Revenues (BUSD)	413	465	510	570	476	564	642	706	720	767	721	709	754
Operating profit (BUSD)	4.4	15.0	19.1	−1.1	1.9	27.6	19.8	18.4	25.3	41.7	59.8	65.2	62.6

Source: IATA Economic Performance of the Airline Industry, December 2017.

cost reduction to a ticket price reduction, moving more passenger with lower costs and reduced revenues. During 2015 and 2016, the fall of the oil price was so deep (2016 average price was 45% of the 2014 price), that operating profits reached record levels, in spite of the revenues descend.

Air transport sector is very active in the energetic efficiency improvement by very good reasons. Historically the fuel bill has been between one fifth and one third of its total expenses. Fuel efficiency, measured in terms of liters of kerosene per RTK has been continuously improving at a rate about 1.5%–2% per year, independently of the oil price fluctuations. However, the fast growth of the traffic (over 4% annually during the last 20 years) makes that the global consumption figure is greater every year.

The huge fluctuations of the oil price might create some doubts about the energy efficiency evolution of the air industry. A typical tradeoff for airline planners is the replacement of some aircraft in operation by new more efficient models. The economic rationale is to recover the money invested in new aircraft thanks to the operation savings coming from advanced technologies incorporated in modern models. If the oil price goes down, the operating cost advantages of the new equipment are reduced and may give an opportunity to keep the old models in service some additional years.

The relationship between the oil price oscillations and the fuel efficiency of the new commercial aircraft models, entering into service, is explored in Fig. 2.9. The efficiency parameter is fuel grams per RPK and its trend appears not having the predicted inverse relation with the average fuel price, in USD per USG. An argument with undoubted weight is the design-to-build cycle of commercial aircraft that is close to 10 years. Therefore, manufacturers work on oil price forecast and not on actual prices.

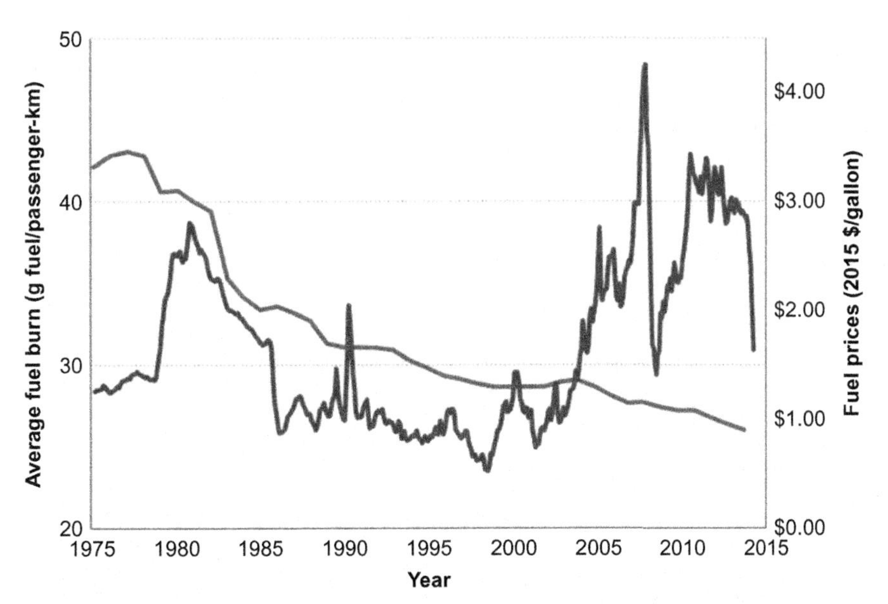

Fig. 2.9 Fuel efficiency of the new commercial aircraft types and evolution of oil price. *Source:* Kharina, A., Rutherford, D., 2015. Fuel Efficiency Trends for New Commercial Jet Aircraft: 1960–2014. ICCT.

On the other hand, airlines are in a similar situation, but with a much shorter timeline: the individual aircraft lifespan is 20-30 years, while a new aircraft model family may be in production 40 years or more.

Airlines have very limited resources to control their fuel budget. Their operation departments work hard into reducing fuel consumption, within the limitations imposed by safety and punctuality and the commercial department requirements. A detailed analysis of all the flight planning and operating procedures to save fuel is in Chapters 5 and 6 of this book.

The airline purchasing and finance groups have the task of negotiate with fuel providers and search the most proficient way of financing their acquisitions. The discussion tools are purchasing volumes in the different points served by the airline flights and the possibility of hedging. The first one is applicable in negotiations of large airlines and fuel suppliers. Local airlines have a good hand with local providers, but in other zones, with low volume of fuel requirement, a procedure being in rising trend is the creation of a group of airlines (generally belonging to one of the largest alliances) to jointly deal with fuel suppliers.

Hedging is the airline version of the future dealings in the stock market. It was created in the 80's, when fuel price was steadily increasing. An airline may buy a certain amount of fuel to be delivered this year or later on in the future, at a fixed price. Then, if oil cost increases more than the agreed price, the airline made a profit, or a loss if the trend goes in the opposite direction.

This practice has caused great discussions among supporters and opponents. For the financial department, it has the advantage of making fuel budget predictable and the possibility of doing some savings depends on how trustable are the oil price short term forecast. A number of airlines rejected flatly the procedure from the beginning but most of them have changed positions in the recent years. It is safe to say that a great majority of the large airlines do edging, but keeping different policies on the amount of their consumption covered by this practice and on the temporal range of their commitments.

2.4 Kerosene and alternative fuels

The name kerosene is usually applied to the aviation fuel burned by aircraft gas turbine engines, either commercial or military, but the kerosene specification may change, depending on the type of application. The fuels for military consumption have a wider specification and are named with the letters JP (jet propellant), followed by a number from 1 to 10. Civil kerosene are Jet A or Jet B families.

Jet B is a mix of kerosene and gasoline in a 30–70 proportion. It is lighter than Jet A and more difficult to handle due to its high flammability feature. It has the advantage of a very low freezing point of −60°C and is similar to military JP-4. In the civil field, it is used in Alaska, Northern Canada and Russia, in relatively small quantities.

The basic commercial service kerosene is the Jet A family. The most common one is Jet A-1, with a typical density of 0.804 kg/L, specific energy 42.8 MJ/kg, flash point 38° C and a freezing point of −47°C. The other variant, known as Jet A only, is slightly heavier with a 0.820 kg/L density, practically same specific energy and flash point and a higher freezing point of −40°C.

The key difference is the freezing point. Commercial aircraft fly at cruise altitudes in which external air temperature may reach −60°C. As the range of the aircraft increases and nonstop flights cover greater distances, the time that aircraft wings spent at such very low temperatures becomes longer and due care should be given to avoid that frozen fuel locks the fuel pumps taken the kerosene to the engines and interrupt the feeding. Jet A freezing point has demonstrated to be too high for long flights over the North Pole and was almost totally replaced by Jet A-1. Today, it is still available in United States and some places in Canada

There have been a high level of research and the consequent testing to find an alternative to kerosene. The reasons are diverse: first, oil is a nonrenewable resource and will be finished at an unknown date in the future; second, having an alternative source might give airlines a choice, and getting some more control on the fuel market than depending of a single provider; and finally, the alternative fuel should be more ecological and its production cycle would leave a smaller carbon footprint, reducing local and climate change emissions.

The task of finding the right product is proving to be extremely difficult. A key element is the amount of changes that a new and different fuel might require in the air transport system. Aircraft and airport design are made based on using gasoline, kerosene or some other fuel of similar chemical and mechanical properties. Departing of this assumption leads to a major refurbishment of the complete logistics (transportation, distribution, storage) and, perhaps, new requirements for the engines and fuel tanks design. As commercial aircraft architecture is a very integrated discipline, such changes suggest a total redesign of the airplane in a different structure than today's tube-with-wings.

The two main research lines in this area are focused on the production of a liquid so similar to the fossil kerosene that can be mixed with it, keeping the same properties. The term drop-in describes these type of fuels. A second and relatively new approach is the use of some electrical source of energy, if not for a total replacement of the kerosene, for a partial substitution of some of the tasks requiring engine energy in modern aircraft, like air conditioning system, or hydraulic systems. A second step would be the hybrid concept, with the aircraft using kerosene in high powered phases of the flight (takeoff, initial climb) and going to electricity in low power ones, like cruise or approach.

In both applications, the immediate advantages would be environmental, if the life cycle of the drop-in fuel is leaving a smaller carbon footprint than fossil kerosene, or the consumed electricity is obtained in a renewable way. The hypothetical energetic advantage is doubtful in the first case, because the aircraft is using exactly the same amount of energy with the same efficiency. Our present experience shows slight heating power advantages in some tested drop-in biofuels, but always of a very small magnitude.

The systematic use of electricity is not enough developed to establish accurate figures. The replacement of hydraulic and pneumatic system by electrical elements would reduce the energy extracted from the engines and, in case this is better than the energy needed for transporting bigger batteries, there will be an improvement in efficiency. As in all the cases of weight versus energy comparisons, the result is more favorable for short and medium range models. The progress is being slow, with some programs initially titled "All Electric Aircraft", moving to the more prudent name of "More Electric Aircraft". Some applications of those technologies are already in commercial service. The Boeing B-787 entered into service in October 2011, using electricity to replace pneumatic power and wing anti-ice systems. The manufacturer declares that fuel savings might go up to 3%, adding up less consumption and lower weight. Some important technical problems with the high power ion-lithium batteries during the initial service seems to indicate that this technology needs still some additional maturation time.

Electric engines to replace turbine ones or hybrid engines, a combination of both propulsive technologies, are relatively new, because the energy/weight ratio of the batteries has not yet reached the values needed for a commercial aircraft engine. Some light models have flown with different power plants: Boeing demonstrated in 2008 a 770 kg. Maximum Takeoff Mass (MTOM), two-seater Diamond aircraft, that flew with a hydrogen fuel cell replacing a 80 hp piston engine. During the 2014–17 period Airbus tested a 550 kg MTOM E-Fan, also a two-seater, with two electric engines fed by Lithium-ion batteries. None of those programs has been continued, but both Airbus and Boeing are supporting additional research towards a hybrid regional airliner in the 50–80 seats category.

The bio-kerosene or drop-in fuels are much more technical defined, based in the automotive long experience with ethanol, corn and soybean feedstock. Many airlines has performed regular flights with a mix of bio and fossil kerosene without any incidence. There is a bio kerosene approved certification (ASTM D7566) since 2011 and modern engines are certified for using a mix of up to 50% drop-in fuel.

The key point is the lack of economic viability. The bio kerosene cost can be in the order of 200–250 USD per equivalent barrel, 3 or 4 times the present price of fossil kerosene. Here is a lot of research on the best feedstock, moving towards oily plants not competing with food production, like jathrofa or camelina, and later on microscopic algae. In any case, the production scale is small and it is expected that the unit cost might improve something with a higher volume production, but it is unlikely to reach a competitive price situation on industrial basis.

Automotive fuels have a mandatory minimum level of bio component inside. The rationale is that the CO_2 savings, in life cycle basis, should have a price. If carbon price is included, a similar regulatory procedure could be applied to bio kerosene in order to obtain a reduction in the aviation sector contribution to climate change. After ratification of the November 2015 Paris Agreement, happened one year later, pressure is increasing to include carbon costs in any economic policy related with energy.

The elements of the air transport system

3.1 Airlines

Airlines are the central element of the air transport system. In order to understand the implications of energy efficiency for airlines, it is interesting to describe their different typologies and working procedures. These typologies are based on the air transport demand structure, which can be segmented in the following way:

- Passengers transport
  - Business travels
  - Leisure travels
  - Visiting friends and relatives (*VFR*) and other purposes not included in the two previous categories
- Cargo transport

When segmenting the passengers' air transport market, the variables to be considered are:

- The purpose of the trip (business, leisure, etc.)
- The trip duration
- The sociocultural characteristics of the traveler
- The specific requirements of the travelers of each segment

In business travels, the client is directly the traveler or, very often, the person that decides the travel policy in the company, and the main concern is the travel success. The airlines offer for short-range business travels is characterized by the high frequency of flights, which are concentrated in the early morning and evening hours. Reliability and punctuality are perceived by passengers as a critical factor. Flexibility with respect to the access to the flight and the intermodal connectivity at the destination airports are also important. Passengers are sensitive to the loyalty programs of the airlines and require a good service at the airport (for instance, appropriate waiting spaces or lounges), as well as a good service in cabin. In exchange, it is passenger providing high-yield (average revenue per PRK) values.

In leisure or tourism travels, the client can be the passenger himself, or a tour operator, for instance, a travel agency. Main concerns in this case are cost savings and the facility to plan the travel. Contrary to the business travels, for tourism travels, there is no demand for specific hours of the flights (morning, afternoon, or evening), although there is a concentration of the demand during certain periods of the year (seasonal demand). Cost savings being a fundamental concern for this passengers, they provide a low value of yield. These passengers do not demand an especially good service in cabin and accept longer check-in times or special departure times like in the middle of the night.

Energy Efficiency in Air Transportation. https://doi.org/10.1016/B978-0-12-812581-6.00003-X

Affinity travels, usually known as Visiting Friends and Relatives (VFR), are a diverse set of travelers, including the so-called ethnic tourism (migrants traveling periodically to their countries of origin), displacements due to special celebrations (Christmas in Europe, Thanks Giving in the United States), travels to sport or shows events, and other motivations. They usually search for cheap fares, buying the tickets very well in advance, but are not flexible with respect to travel dates.

Finally, regarding cargo transport, the client is the person or organization needing a transport service whose requirements can only be met by air transport. The main concerns are the fast delivery and the possibility to overcome geographical obstacles. The service can be an airport-to-airport transport, provided by a cargo airline, or a door-to-door transport, normally multimodal, performed by an integrator such as Federal Express or UPS.

This segmentation of the demand is fundamental for the airlines yield management, trying to maximize the revenues of the company by controlling the optimal mix of fares.

The strategy of airlines has been evolving, getting adapted to the previously described different types of demand, and also to the liberalization processes that have been (and still are) taking place in different regions of the world, especially in the United States and in Europe. Attending to their degree of diversification, i.e., the number of different products they offer, and their fares, the airlines can adopt any of the strategies shown in Fig. 3.1.

As a result, airlines can be classified in the following way:

- Passengers transport
  - Network carriers
  - Low-cost operators
  - Charter operators
  - Regional airlines
- Cargo transport
  - Cargo operators
  - Integrators

Network carriers, sometimes also named traditional or incumbent airlines because in many cases they correspond to the old flagship airlines of different countries prior to the air transport liberalization processes, present the following main features:

- Hub & spoke strategy, allowing them to offer a diversified routes network, concentrated in one or several hub airports (distribution center), basing their traffic in a high number of connecting passengers.
- Maximization of the number of markets they serve.
- Because of the routes variety, operation of several different aircraft types, with diverse capacities and ranges.

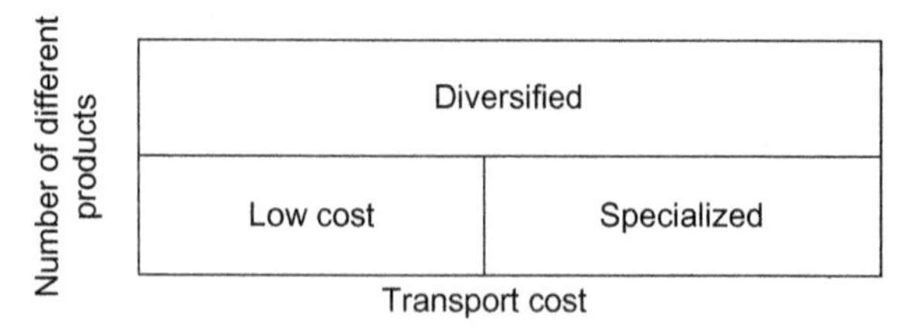

Fig. 3.1 Different strategies of airlines.

- Strong dependency on the infrastructure so that the connection model works.
- Multiproduct strategy, with various classes in cabin (First Class, Business Class, Tourist or Economy, etc.), corresponding to different levels of service to passengers.
- Strong sophistication in First and Business classes as a way of differentiation among airlines.
- Large fares variety.
- Passengers loyalty programs (Frequent Flyer Programs)
- Membership to an Airline Alliance.
- High volume of sales through travel agencies
- Utilization of Global Distribution Systems

Low-cost companies compete in fares and frequencies in mainly short- and medium-range routes, with point-to-point traffic, offering very little different fares, sold primarily on Internet, providing minimal service to the passenger at a very low fare, and obtaining additional income through the so-called ancillary revenues associated to added services such as luggage check-in, seat choice, etc. In order for this business model to work, these companies must keep a minimum cost, reducing at maximum in all areas of activity. The following list is indicative of cost reduction measures:

- Operation in secondary airports, with the corresponding reduction in airport charges.
- Direct sales, primarily on internet or by call centers, without agencies
- Reducing the outsourcing of areas like handling or maintenance, keeping always control on those costs
- Digitalization of all working information and data, limiting the utilization of paper
- Low labor costs

In addition, in order to reduce their unit cost, these companies work on improving the utilization of their assets, by minimizing the variety of aircraft they operate (generally all their aircraft correspond to the same model or belong to the same aircraft family). They try to utilize those aircraft more hours a day, shortening their turnaround times (which is possible because they do not have connections or cargo services).

Although those aforementioned features are typical of the so-called Low-Cost Carriers, there is a great variety of companies mixing them with features attributed to traditional carriers, like starting point-to-point long-range routes or having a small business class. At the same time, the network carriers are introducing some of the low-cost airlines tactics, like offering low fares without additional services (seat selection or baggage allowance), leaving the connection offer as the most relevant differential feature.

Charter companies appeared first in Europe, due to the restrictive legislation existing until 1993. They approach just a single market segment, the tourism travel, and base their strategy in the sale of packages of seats to tour operators and travel agencies. The latter are the ones who sell the tickets to the passengers, normally within an integrated vacation package (including hotel, entertainment, and other services at the destination). Different from scheduled flights airlines, charter companies offer flights on demand. For this reason, their load factors are usually large, and their fares (the part of the vacation package corresponding to the flight) considerably smaller than those of the scheduled flights. The eruption of low-cost companies put charter companies in a very difficult situation, with a trend to get vertically integrated with hotel companies and tour operators.

Regional airlines are specialized in passenger transport in generally short routes and because of that, very often, in domestic flights (or intracommunitarian in the European case). They operate fleets of regional aircraft, with less than 100 seats. Although some of them still operate independently, most of them operate as franchisees or with some type of agreement with network carriers, which use them to supply passengers to their long-range flights.

Regarding cargo transport, there are major differences between passenger and cargo transport, leading to the specialization of airlines (and airports as well). Specific features of air cargo transport are:

- Strong unidirectionality
- High weight-to-price ratio
- Need of larger handling and stocking resources
- Very heterogeneous payload
- Huge competition with ground transport (boat, train, trucks)
- Good safety and security compared with the other transportation modes

The three large segments of cargo transportation are:

- Emergency transport
- Perishable goods
- Logistic transport

Integrators are large multimodal companies, which are specialized in gate-to-gate deliveries within a maximum specified time. Their aeronautical divisions are very large and normally use hub & spoke networks for the distribution of their deliveries. Globally the major ones are Federal Express, UPS, and DHL.

The market share of the world 200 largest airlines is shown in Table 3.1 classified by company type and in Table 3.2 organized by world region.

Table 3.1 Market share of the world 200 largest airlines classified by company type (2016 data)

Airline type	Market share (% PKT total)	L/F (%)	Growth (% PKT)
Charter	3.3	87.3	3.9
Low cost	19.4	81.4	9.0
Network carriers	75.8	79.6	5.0
Regionals	1.5	78.3	3.6

Table 3.2 Market share of the world 200 largest airlines organized by world region (2016 data)

Region	Market share (% PKT total)	L/F (%)	Growth (% PKT)
Africa	1.4	69.9	2.0
Asia-Pacific	31.2	78.3	7.7
Europe	27.4	81.6	5.5
Middle East	9.8	77.3	10.6
North America	25.9	83.0	1.9
Latin America	4.3	78.2	5.7

3.2 Aircraft

As per July 2017, more than 28,000 commercial transport aircraft were registered as being in operative status. However about 700 of them may be grounded, not performing commercial flights, but still not taken out of the National Registries. Their geographical distribution is shown in Fig. 3.2, and the repartition per aircraft manufacturer is shown in Table 3.3.

Concerning aircraft manufacturers, Boeing still leads in terms of operative aircraft; with more than 11,000 units compared to the almost 9000 units of Airbus (aircraft inherited from McDonnell Douglas are included in the Boeing's figure). There are some 4700 operative regional jets, with Embraer heading the list with almost 1700 units followed by Bombardier with 1300. With respect to turboprops, there are about 5000, the Dash-8 model with 880 units being the most popular, followed by the ATR72 with 950. It is interesting to note that out of those 5000 operating turboprops, around 500 manufactured in Russia, most of them flying in Russian airlines or in airlines from related countries, like Cuba (Cubana, Aerogaviota) or North Korea (Air Koryo). The share for Russian jets is about 750 in operation, distributed in Russian companies, compared to more than 20,000 from the large western manufacturers.

In general terms, the market of commercial transport aircraft can be divided, according to the aircraft capacity and range, into the following segments (aircraft seating represents high densities in each model):

- Medium/long-range jets (more than 120 seats)
- Regional jets (between 30 and 120 seats)
- Regional turboprops (less than 90 seats)

Medium/long-range aircraft can be divided into single-aisle models (narrow body) and double-aisle models (wide body). Aircraft manufacturers are specialized according

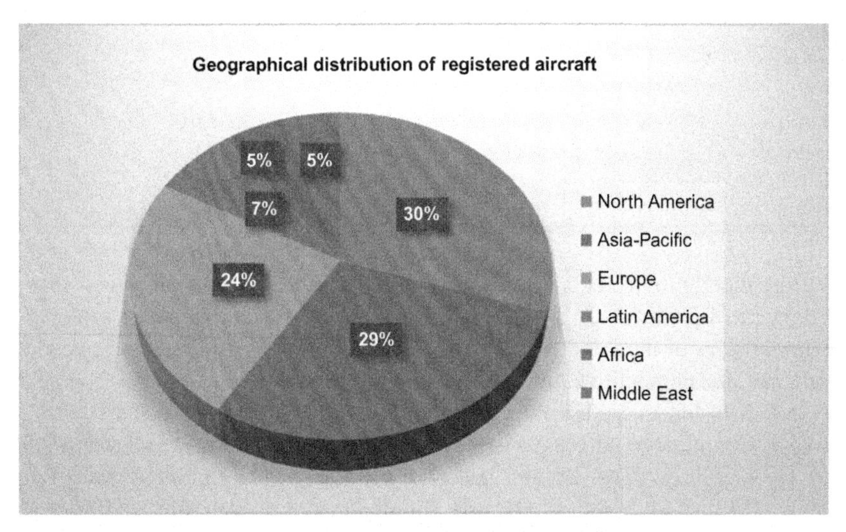

Fig. 3.2 Geographical distribution of registered aircraft (*Ascend fleets July 2017*)

Table 3.3 **Distribution of registered aircraft per aircraft manufacturer (ACAS database, July 2017)**

Manufacturer	Fleet	(%)	Δ% 17/16
Boeing/MDC[a]	11,564	41.0	+4.2
Airbus[a]	8915	31.6	+7.3
Bombardier[a]	2173	7.7	−0.7
Embraer[a]	1783	7.0	−2.1
ATR[a]	950	6.3	+6.3
HawkerBeech	435	1.5	−4.3
Antonov[a]	353	1.3	+0.1
BAE Systems	320	1.1	−4.1
SAAB	258	0.9	−3.0
Fokker	222	0.8	−6.9
Other manufacturers[a]	1262	4.5	−16.2
Total	28,235	100	+3.6

[a] Manufacturers with commercial models in production

Table 3.4 **List of world aircraft manufacturers, with the type of aircraft they produce and the range of seats**

Company	Type of aircraft	Seats
Airbus	Jets > 100 seats	120–650
Boeing	Jets > 100 seats	120–500
Ilyushin	Jets	250–350
	Turboprops	50–70
Tupolev	Jets	100–200
UAC	Jets	120–200
COMAC	Jets and Turboprops	60–200
Embraer	Regional Jets	40–120
Bombardier	Regional Jets and Turboprops	40–140
Sukhoi	Regional Jets	70–120
Mitsubishi	Regional Jets	70–90
ATR	Turbo props	40–80
Antonov	Regional Jets and Cargo	50–70

to the just-mentioned categories. The list of world aircraft manufacturers, together with the type of aircraft they produce and the range of seats, is shown in Table 3.4.

After decades of market concentration, and due to the large investment needed to develop new and competitive products, there are a limited number of aircraft manufacturers in the world. A very brief description of them is provided hereafter.

Boeing Commercial Airplanes is part of The Boeing Company and was founded in the USA in July 1916. Boeing has final assembly lines in Everett and Renton (Washington) and in North Charleston (South Carolina). An important milestone was the acquisition of McDonnell-Douglas in 1997 (with 3387 commercial jets manufactured to date). The cumulated number of orders and deliveries are approximately 25,000 and 20,000, respectively. The Boeing product portfolio consists of the following models:

- Single aisle: B737 Series 700, 800, and 900 (100–200 seats) in production, including the executive version BBJ and the initial models of the new B737MAX family.
- Double aisle: B747, B767, B787-8 and -9, B777 (190–480 seats) in production (cargo and passengers).
- B707, B717, B727, B737 (Series 100, 200, 300, 400, 500, 600), B757, DC8, DC9, DC10, MD11, MD80, and MD90 out of production (cargo and passengers).
- In development: members of the B737MAX family, B787-10, and B777-8, and -9.

Boeing's big competitor is Airbus, founded in 1970 merging a large part of the aerospace industry of Germany, France, UK, and Spain. Today Airbus has four final assembly lines for their commercial aircraft, two in Europe (Toulouse and Hamburg), one in China (Tianjing), and the latest one in the USA (Mobile, Alabama). As per January 2018, Airbus had received a total of 18,206 orders and delivered 10,953 aircraft. Airbus product portfolio includes commercial jets in the range between 120 and 650 seats and military transport aircraft. In the commercial side, Airbus products are:

- Single-aisle aircraft: A318, 319, 320 (ceo and neo) and A321 (between 100 and 220 seats), together with executive versions of A318, A319, and A320.
- Double aisle: A330 Series 200 and 300 (220–270 seats) and cargo versions, A350-900 (270–320 seats) A380-800 (450–650 plazas). Still in service but out of production there are the following double-aisle aircraft: A300 and A310 (passenger and cargo versions), A340 Series 200, 300, 500, and 600 (250–370 seats).
- The products, which are in development, are: A350-1000, A330-800 and 900, and versions of A319 and 321 (neo) and cargo.

On the side of smaller aircraft manufacturers, Bombardier and Embraer hold a quasiduopoly status. Bombardier is a Canadian industrial conglomerate that entered the aerospace sector in 1986 buying different manufacturers since then like Canadair, Shorts, Learjet, and De Havilland. Bombardier produces both Turboprops and Regional Jets. Concerning Turboprops, the model in production is the Dash8-400 (80 seats). Out of production are the Dash8-100, -200 and -300, Dash-6 Twin Otter (new version from Viking), and Dash-7. The overall orders and deliveries figures at the end of 2017 for the Dash8 are 1285/1242. On the Regional Jets side, the following models are in production:

- CRJ700, CRJ900, and CRJ1000 (65–100 seats).
  The smaller versions of the family, the CRJ100 and CRJ200 (50 seats) are already out of production. Total number of orders and deliveries of this Regional Jets family are 1918 and 1876, respectively.
- C-Series 100 and 300 (100–160 seats).
  This new aircraft family is launched to compete with the smallest Airbus and Boeing family. In January 2018, they had got 372 sales and 24 deliveries.

The "Empresa Brasileira de Aeronautica," Embraer, was founded in 1969 to develop both civil and military programs. Their commercial jets offer includes the Regional Jets family E-Jet-170/175/190/195 (65–120 seats) in production. Already out of production are the Regional Jets ERJ-135/140/145 and the Turboprops EMB-110 and 130. Embraer have the E2 Regional Jets family in development with capacity for 75–130 seats. Overall, Embraer have received orders for 2636 Regional Jets and have delivered 2157.

The turboprops of Bombardier compete with *Avions de Transport Régional* (ATR) was founded in 1982 as a joint venture between the French Aerospatiale (today Airbus) and the Italian Alenia-Finmeccanica (today Leonardo), with a participation of 50% each. Based in Toulouse, ATR produces the Turboprop family ATR42 (50 seats) and ATR72 (78 seats), having received 1538 orders and delivered 1300 aircraft.

In the old Soviet Union, a number of manufacturers developed aircraft in the different market segments. Ilyushin was founded in the Soviet Union in 1932. In 1992, the Il-96 entered into service, with two versions, the series 300, with 250 seats, and the 400T cargo version with 92-ton capacity. The total number of orders and deliveries is 37 and 36, respectively. Ilyushin also offers the Turboprop Il-114 Series 100 and 300 (50–70 seats) with 20 civil aircraft sold and delivered.

Founded in 1922 in the Soviet Union also, Tupolev models in production are the Tu-204/214 (medium range, 200 seats) and Tu-204/300 (long range, 200 seats) that entered into service in 1996. Tupolev has received 91 orders and delivered 76 aircraft. Models already out of production are the Tu-154 and Tu-134.

Sukhoi was founded in 1930 in the Soviet Union. Sukhoi is producing today the Regional Jet Superjet SSJ100, in cooperation with the Italian Finmeccanica and engines from the French SNECMA. The SSJ100 was certified in 2010 for the passenger version and a business version, SBJ, was certified in 2014. Sukhoi has received 179 orders of the SSJ100 and delivered 94 aircraft.

Antonov was founded in the Soviet Union in 1946. The product portfolio of Antonov includes:

- Regional Jet family An-140/148/158/178 (70–85 seats), produced in Ukraine and Russia, with 26 orders and 15 deliveries (civilian versions)
- Cargo aircraft An-124 and An-225 (100 ton), with orders and deliveries of 22/22 and 2/1, respectively, for civilian versions.

The old Soviet industry is being reorganized around UAC. The United Aircraft Corporation of Russia was founded in 2009, as the core for the design and manufacturing of the future Russian commercial jets. UAC coordinates different design offices and manufacturers around Ilyushin. UAC is developing the MS-21 Series 200, 300, and 400 with a scheduled initial entry into service in 2020 and 257 orders so far. Number of net orders may be debatable because it is not very transparent and it is difficult to differentiate orders from operators and from financial institutions.

In China, the Commercial Aircraft Corporation of China Ltd. (COMAC) was founded in 2002 as a division of AVIC, the China State Aeronautical Company. Aircraft in production are:

- Regional Jet ARJ21-700 (70–90 seats) with first deliveries in 2015 and 302 orders and 2 deliveries.
- Turboprop MA 60 (60 seats) with 181 orders and 88 deliveries

COMAC is developing the C-919 (140–200 seats) and MA 700 (70–80 seats) with scheduled first deliveries in 2020 and 2019, respectively. The orders are 245 for the C-919 and 203 for the MA 700. Very recently, Russian and Chinese governments have signed an agreement to jointly develop a 250-seater long-range aircraft with a temporal horizon of 2025.

Finally, Mitsubishi was founded in Japan in 1928 as part of Mitsubishi Heavy Industries. It has the Regional Jet MRJ70-90 (70–90 seats) in development, with 233 orders and first deliveries scheduled in 2020.

As mentioned before, not all these aircraft manufacturers compete in the same market segments; they are normally specialized in one or two of the categories in which commercial aircraft can be classified according to their capacity and range. In the segment of aircraft with more than 120 seats Airbus and Boeing have been maintaining for years a duopoly regime (Fig. 3.3), sharing the market and covering all potential needs spectrum (Table 3.5).

Aircraft in this segment from manufacturers from China and Russia (or the Soviet Union before) have had traditionally a very low level of deliveries (as it has been seen

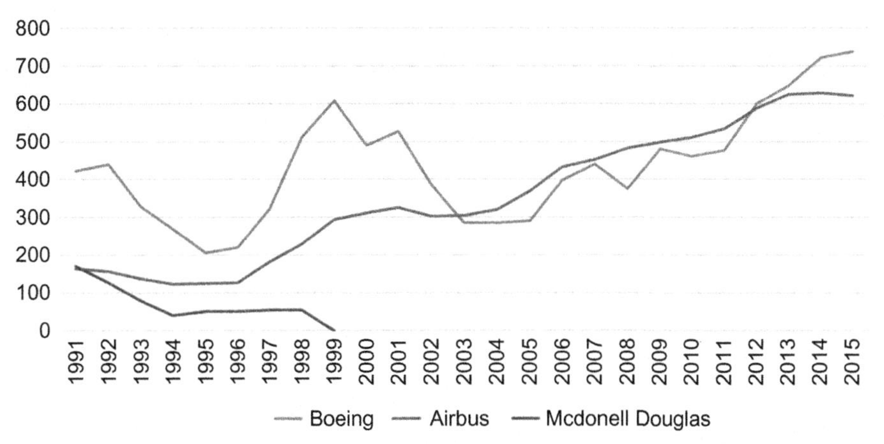

Fig. 3.3 Commercial aircraft deliveries (aircraft with more than 120 seats; Eastern Europe and China are not included).

Table 3.5 Product portfolio of Airbus and Boeing

Company	Models	Seats
Airbus	A318/319/320/321	100–220
	A330	220–270
	A350	200–400
	A380	450–650
Boeing	B737	100–200
	B787	200–300
	B777	300–400
	B747	400–550

in the previous section) and have not been a true competition to western manufacturers, with their sales being restricted to their respective domestic markets and those under their political influence.

It must be noticed that analyzing the passengers' world traffic evolution compared to the number of orders and deliveries of the aircraft manufacturers, it is observed that while the orders follow normally the traffic evolution (they both grow in the same periods of time), the deliveries do not (they grow in different periods). The reason is that aircraft manufacturers plan their deliveries to every airline (slots) and commit them well in advance due to the complexity of the manufacturing process of an aircraft. This explains partly the lessors business, because they have a kind of stock of aircraft, which are available to the airlines in a shorter time than directly through the aircraft manufacturer. Ultimately, the gap between deliveries and traffic evolution is an issue for airlines because sometimes the deliveries arrive in periods of smaller demand for the airline and they have to cancel orders or find another buyer for the surplus aircraft.

The size of the fleet in service in July 2017 of the Airbus and Boeing aircraft is shown in Table 3.6.

A similar situation exists in the regional jets segment, with the Canadian manufacturer Bombardier and the Brazilian company Embraer sharing almost equally the market of jets between 30 and 100 seats (Fig. 3.4 and Tables 3.7 and 3.8).

A third aircraft market segment is that of Turboprops with capacities in the range between 50 and 90 seats approximately for short routes. After some years of very poor sales, this market has been reactivated because of the fuel price increase, as turboprops are much more fuel efficient than their jet counterparts. The existing programs can be seen in Table 3.9, with the market split between Bombardier and the ATR consortium. In this segment, the potential of Chinese, Russian, and Ukrainian manufacturers is currently difficult to forecast because although there some turboprops already in operation from AVIC, Ilyushin, and Antonov, their respective civil version productions are very limited and the exporting sales are almost nonexistent.

Table 3.6 **Airbus and Boeing fleet in July 2017 (ACAS database)**

	Model	Fleet	Δ% 17/16
1	Airbus A320 ceo family[a]	6838	+4.9
2	Boeing B737NG[a]	5968	+7.4
3	Boeing B777[a]	1387	+5.2
4	Airbus A330[a]	1214	+3.8
5	Boeing B737 2nd gen.	890	−4.4
6	Boeing B767[a]	744	+0.8
7	Boeing B757	689	+0.1
8	Boeing B717/MD80-90/DC9	607	−7.0
9	Boeing B787[a]	554	+31.3
10	Boeing B747[a]	489	−2.8

[a] Models in production

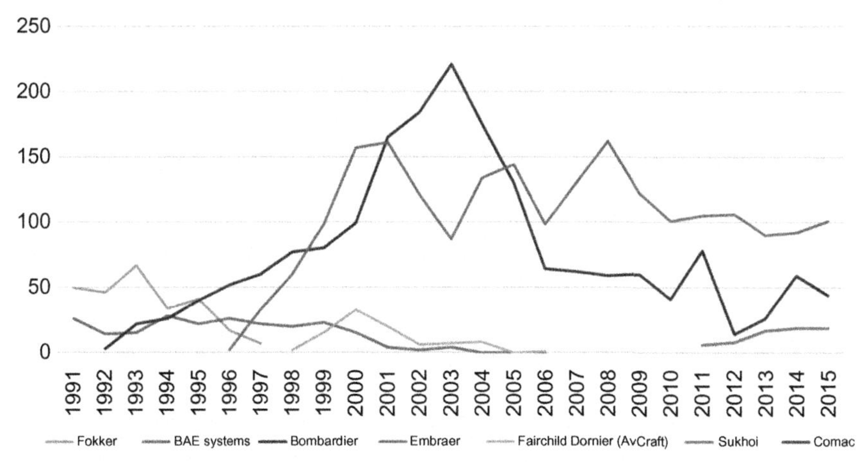

Fig. 3.4 Commercial aircraft deliveries (regional jets with 30–120 seats; Eastern Europe and China are not included before 2010).

Table 3.7 Product portfolio of Bombardier and Embraer

Company	Models	Seats
Bombardier	CRJ200/700/900/1000	50–95
Embraer	ERJ135/140/145	35–50
	E170/175/190/195	70–115

Table 3.8 Size of the fleet of Bombardier and Embraer by July 2017 (ACAS database)

	Model	Fleet	Δ% 17/16
1	Embraer E-Jet family[a]	1235	+8.3
2	ATR 42/72[a]	950	+4.1
3	Bombardier Dash8[a]	880	+2.0
4	Bombardier CRJ700/900/100[a]	762	+7.6
5	Bombardier CRJ100/200	516	−7.4
6	Embraer ERJ135/140/145	454	−14.0
7	Beechcraft 1900	328	−3.0
8	De Havilland Twin Otter[a]	270	+1.5
9	Saab 340	225	−2.6
10	Fairchild Metro/Merlin	219	+3.5

[a] Models in production.

Table 3.9 Product portfolio of Turboprops

Company	Models	Seats
BOMBARDIER	Dash8Q	45–90
ATR	ATR42/72	50/75
AVIC	MA60	50–60

Table 3.10 **Engine types per manufacturer**

Company	Engine types	Families (max. thrust in pounds)	Applications
General Electric	Turbofan and Turboprop	GEnx (53,000–75,000) GE90 (74,000–115,000) CF6 (40,000–72,000) CF34 (9220–20,000)	B787, B747, B777-8 and -9 B777 B747, B767, A330, DC10, MD11, A300, A310 CRJ, E-Jet, ARJ21
Rolls-Royce	Turbofan and Turboprop	Trent (53,000–95,000) RB211 AE3007	B787, A350, A380, A340, B777, A330 B747, Tu-204, L1011, B757 ERJ
Pratt & Whitney	Turbofan and Turboprop	PW4000 (52,000–98,000) PW2000 PW6000 (22,100–23,800) PW1000G (22,100–23,800) PW100	B747, B767, A330, B777, A310 Il-96, B757 A318 C-Series, MRJ, MS-21, A320NEO, E2 ATR, Dash-8
Russian	Turbofan and Turboprop	Ivchenko D-436 PERM PS90	Tu-334 IL-96, Tu-200

As is very well known, engines are designed and produced by companies different from those that develop aircraft, and both are a completely different market. There are three big turbofan manufacturers in the market of engines for commercial aircraft: General Electric, Pratt & Whitney, and Rolls-Royce. Sometimes, these three manufacturers make up consortia to address the development of a particular engine. Examples of these consortia are CFM International (General Electric and the French SNECMA), International Aero Engines, IAE (Pratt & Whitney, Rolls-Royce, the German MTU, and the Japanese Japan Aero Engines), and the timely association between General Electric and Pratt & Whitney for the development of an engine for the Airbus A380. Alone or in consortia, these three manufacturers share the commercial jet engines market. The most representative engines of each manufacturer are shown in Table 3.10, together with the aircraft, which are equipped with them.

The products developed by each consortium are:

- CFM International (GE and SNECMA: CFM56 engines for the A320 and B737 families, new motorization of DC8 and B707, A340 Series 300 and 200. The certificated thrust is in the range 18,000–34,000 pounds. The latest version of the CFM56 is the LEAPX, equipping the A320 neo, B737 MAX, and C919.
- International Aero Engines (RR, PW, MTU, JAEC) V2500 engines for the A320 family (with the exception of the A318) with a maximum thrust in the range 23,000–32,000 pounds.
- Engine Alliance (PW and GE) for the A380 family (GP7000, 76,500–81,500 pounds)
- SaM (SNECMA and NPO Saturn) for the engines in the Sukhoi SuperJet

Like in the case of the aircraft design and manufacturing, an initial engine version is developed along the time producing the different variants that are in the market today, much more advanced due to the continuous incorporation of technological advances. For instance, the Trent 500 engine from Rolls-Royce, which is installed in the Airbus A340-600, is a Rolls Royce engine designed in the mid-1990s, and it is a derivative of the RB-211 engine designed in the 1970s.

The situation of market repartition between two players, Airbus and Boeing on the medium-long range jets and Bombardier and Embraer on the regional jets, is changing in the last years. On one side, both Bombardier and Embraer have projects for jets with more than 100 seats. In fact, Embraer with its E195E2 will be competing somehow with the smaller models of the A320 family, and the B737 family. On the other side, Bombardier has launched its C series family with capacity in the range between 100 and 150 seats.

But, as mentioned at the beginning of this section, there are also manufacturers from China and Russia entering this market. Both COMAC (Commercial Aircraft Corporation) from China with its C919 (150–200 seats) and United Aircraft Corporation of Russia with its MS21 model (150–210 seats) are trying to access the market with their respective models and compete directly with the narrow body aircraft of Airbus and Boeing—the A320 and B737 families, respectively. Estimated entry into service date for the C919 and MS21 is 2020 in both cases, once they are able to overpass the difficulties of the complex modern commercial certification.

The IRKUT Corporation was established in 2002 based on Irkutsk Aviation Production Association. In November 2006, the IRKUT Corporation became a member of the Russian United Aircraft Corporation (UAC). The IRKUT Corporation develops their R&D activities through the subsidiary company Yakovlev Design Bureau. The main program of the Russian aviation industry is the development and manufacturing of the MC-21 single-aisle airliner family. By an executive order of the President of the Russian Federation in 2010, the IRKUT Corporation was appointed the sole executor of the government contracts for the MC-21 airliner family development and certification.

The MS-21 family comprises MS-21-300 (seating 160–220 pax) and MS-21-200 (seating 130–165 pax). According to the IRKUT Corporation information, the MS-21 competitive advantages are:

- Cash Operating Costs savings with 12%–15% advantages over the aircraft of current generation
- Extended operational capabilities
- The best technical characteristics due to the advanced aerodynamic, use of modern materials, and new-generation engines
- High level of commonality within the family
- Optimal fuselage diameter matching the needs of different types of carriers: from low cost to legacy
- Enhanced level of comfort in the passengers cabin
- Reduced turnaround time
- Meeting future environmental requirements

The MS-21 is intended to be powered initially by the Pratt & Whitney PW1000G and a version powered by the Russian Aviadvigatel PD-14 turbofans will be offered later.

The C919 aircraft is being developed by COMAC, with the purpose of becoming a competitive single-aisle aircraft with a short-to-medium range. The C919 family will cover the 140–200 seat range. According to the company's information, the C919 aircraft complies with advanced design standard integrated high-class powerplant and airborne systems, with excellent flight performance. Two versions are initially envisaged, the C919 ST and the C919 ER. The C919 is equipped with the new LEAP-1C engine developed by CFM as a successor of the CFM56 engine.

But almost simultaneously to those Russian and Chinese developments, in the last years both Airbus and Boeing have introduced improvements in their respective medium-range models, spurred by the high price of fuel and the consequent negative impact on the costs of airlines. These enhancements have been introduced in the already-existing A320 and B737 becoming A320 neo and B737 MAX, respectively. The purpose in both cases is to incorporate to these aircraft the latest technology currently available with respect to aerodynamics, materials (aiming at both reducing structure weight and facilitating repairs) and avionics (following the latest FAA and EASA regulations, such as the Future Air Navigation System (FANS). Nonetheless, the main objective is the reduction of fuel consumption decreasing the operating costs for airlines. The engines that thrust these aircraft are largely responsible for the fuel consumption decrease, as well as the consequent decrease in emissions and in emitted noise as well. These new engines are the CFM International LEAP (for the A320 neo and B737 MAX) and the Pratt & Whitney PW1000G PurePower (A320 neo). Additionally, these new aircraft are capable of transporting a bit more payload and increasing slightly the range with respect to their predecessors. This last feature can make them suitable for transatlantic crossings, allowing low-cost airlines to offer new long-range services to small cities and not only to big metropolis.

As a consequence, some new aircraft types have been recently certified and introduced into service (B787-8 and ATR72-600 in 2012, A350-900 and B787-9 in 2015, A320Neo and Bombardier CS100/300 in 2016). In the 2017–22 period, a bunch of new models will join, like Boeing B737MAX, B787-10, B777-9, and B777-8, other variants of the A320Neo family, A350-1000, A330-800, and A330-900, Embraer E2 family, the new Russian MS-21 and Chinese C919 airliners and new turboprops from Bombardier and Leonardo (former Finmeccanica).

In the other segment, corresponding to regional jets with less than 100 seats, there are also models from other manufacturers that have recently entered into service, like the Sukhoi SSJ, or are in development like the ARJ21 from AVIC in China or the Japanese MRJ70 from Mitsubishi. These new models may break in the medium term the current duopoly of Bombardier and Embraer.

ARJ21 is a new regional jet with 78–90 seats developed by COMAC, equipped with CF34-10A engines. There are two versions: the ARJ21-700 ST, with 1200-nm range, and the ARJ21-700 ER, with a range of 2000 nm. The basic version is already in service for Chinese domestic routes.

The SSJ100 is a new regional jet developed by the Russian manufacturer SUKHOI, with capacity for 98 passengers in two versions: SSJ100/95B, with a maximum range of 3048 km and the SSJ100/95LR, with a maximum range of 4578 km. The SSJ100 is equipped with the new SaM146 advanced turbofan engine, produced by PowerJet, a 50/50 joint company between Snecma of France and NPO Saturn of Russia.

On the other side, it is interesting to observe how many of the aircraft models that today are still being manufactured and delivered are evolutions or derivatives of aircraft that keep the design of decades ago. Nonetheless, improvements in avionics and far more efficient engines are being incorporated. In addition, the manufacturers are gradually increasing the size and capacity of their products in order to accomplish the various needs of airlines. This is notably the case of the Boeing B737, whose first subseries, the B737-100 is dated in 1968 but has today several versions with different capacities (for instance the most popular B737-700 with 149 seats or the B737-900 with up to 220 seats). Another outstanding example is the recent B747-8, which is the latest evolution of the many that the venerable B747, with a first flight back in 1969, has had along all these decades.

The same strategy applies to the European manufacturers, like Airbus: almost 30 years separate the first flights of the A320 and the new A320 neo. Or even ATR, whose two only products (ATR42 and ATR72) are almost identical except in capacity, and their first fight was in the mid 1980.

Boeing and Airbus have been making more difficult the entrance of other manufacturers to this segment by developing the so-called aircraft families, whose commonality levels with respect to flight crews and systems make it more profitable for airlines to incorporate similar models from the same manufacturer. The advantages for the airline operating one of these families can be summarized as:

- Scale savings by the increased number of similar aircraft
- Easy to change aircraft (because of a demand change) in the short term
- Savings in crew costs: less crews, reduced training periods, easier to exchange crews among flights, reduced needs for reserve crews.
- Savings in maintenance due to reduced training courses and less spare parts.
- Saving in Engineering, due to less different Bulletins issued.

The concept shows some drawbacks for the airlines though. The general design is optimized for the first member of the family, not for the others. There is also more vulnerability on the airlines side with respect to technical problems, potentially affecting to a larger number of aircraft, together with the dependency upon a single aircraft (and engine) manufacturer. Finally, there might be renewal problems if the new aircraft family does not cover the same range of performance.

Not only Airbus and Boeing, also Embraer and Bombardier have developed their own aircraft families, as it is shown in Table 3.11.

Concerning future projects, looking at the ongoing research programs, it is possible that in the intermediate future some new products would be launched. Promising candidates are the following:

For aircraft:

- Long-range aircraft:
  - A380 derivative with updated technology (in service date: 2021–2013)
  - Potential Russian-Chinese model with 250 seats (2024–2025)
- Medium-range aircraft:
  - "Middle-of-the-market" 250 seats, 5000-km range (2024–2025)
  - A320 and B737 replacement (2030–35)
  - 100-seat Turboprop (2022–2023)

Table 3.11 Different aircraft families

Manufacturer	Aircraft family	Seats
Airbus	A320 (ceo and neo)	100–240
	A330/340/350	200–450
	A380	450–650
Boeing	B737 (NG and MAX)	100–200
	B787	200–350
	B777	300–400
	B747	400–550
Embraer	E-Jet and E2	65–120
	ERJ	35–50
Bombardier	CRJ	50–100
	C-Series	100–140

For engines:

- Engines with 20,000–40,000 pounds maximum thrust
  - Propfan configuration
  - Fairing Counter-rotating Fan
- Engines with 45,000–110,000 pounds maximum thrust
  - Gear fan
- Engines with less than 20,000 pounds maximum thrust
  - Advanced gear fan
- Turboprops with new blade design

To conclude this section on aircraft and engines manufacturers, it is worth mentioning another important industry intermediating between airlines and aircraft manufacturers: leasing companies. Airlines usually choose between buying aircraft (directly to the manufacturer) and acquiring them through a financial broker or lessor, who is the actual owner of the aircraft and lease it to the operator. In exchange the operator pays a fee. In this latter case, the operator's financial costs are normally reduced because the upfront payment is much smaller than in case of purchasing the aircraft. In addition, it provides more flexibility to the operator in case of needing more aircraft to be incorporated quickly, or in case of having to reduce the fleet size. For the aircraft manufacturers, these leasing companies are generally beneficial because they are customers with firm orders. On the other hand, some of the drawbacks of this formula for the airlines are:

- The rigorous technical conditions for the returning of the aircraft after the leasing period
- Aircraft are not necessarily specified with the technical and commercial requirements of the airline, with the consequent costs for modifications and adaptations
- If the leasing contract terms and conditions are not carefully negotiated by the operator, the financial advantage may be lost and become an over cost.

In 2017, leasing companies owned about one-third of the world commercial aircraft fleet. The top five companies, in term of aircraft value were Aer Cap Holdings, GECAS, Air Lease Corporation, SMBC Aviation Capital, and BOC Aviation Limited.

3.3 Airports

Airports were originally conceived as final terminals of a point-to-point service, close to a big city. Along the years, their installations were getting further from the urban conglomerate they served as the airport space needs grew and its environmental impact increased. Airports became a modal interchange center when the origin-destination journey required a combination of air, ground, and even maritime transport modes.

The liberalization of the air transport commercial aspects introduced new commercial strategies for airlines, increasing the number of connecting flights, mainly due to the configuration of hub & spoke type networks. The main airlines established connecting center at their main bases (hubs) and the fundamental role of the large airport evolved toward connecting flights, with a big percentage of transit passengers with origin and destination in other regions.

The modern concept of the airport tends to an economic activity complex, where the purely transportation function is combined with the generation of economic and commercial projects, acting as a socioeconomic development engine with strong implications in the local territorial organization. As a demonstration of this trend, the nonaeronautical revenues are constantly increasing their contribution to the total revenues of the airport, and in some cases they are even bigger than the revenues coming directly from the aircraft and passengers movement.

The World's airport network is the result of the evolution of the air transport services demand, in parallel to the economic development of the different world regions and the technological advances that allow offering those services with constantly better safety, regularity, and economy conditions.

The principal international passenger traffic flows are shown in Fig. 3.5 corresponding to the year 2015. It is interesting to observe that the number of international passengers in the Asian market is very close to the European one.

This traffic distribution is backed by the classification of the largest world airports by passenger traffic (Table 3.12). Four out of the twelve largest airports are North American, three are Europeans, and five are Asian. The first airport in the list from the Africa/Middle East region is Dubai, hub of Emirates, one of the airlines with a faster expansion, and occupies the position number three. In Latin America, the biggest airport is Mexico D.F, in position number forty three.

With the only exception of Los Angeles, the main entrance point to the United States from the Asia/Pacific region, the rest of the airports in the top twelve list shown in Table 3.12 are the main base of one or several airlines that have built there a hub & spoke system. This starts with Delta in Atlanta, American and United in Chicago, British Airways in London Heathrow, Japan Airlines and All Nippon Airways in Tokyo Haneda, American in Dallas, Air France in Paris Charles de Gaulle, Lufthansa in Frankfurt, Air China in Beijing, United in Denver, Cathay Pacific in Hong Kong, and KLM in Amsterdam. The same consideration applies to Dubai with Emirates, Iberia in Madrid, and Mexico with Aeromexico.

The airports with the fastest growth are mostly from Asia (Table 3.13), concentrated in some of the largest emerging markets: China, India, Korea, and Thailand. With the exception of the airport of La Guardia in New York none of the developed

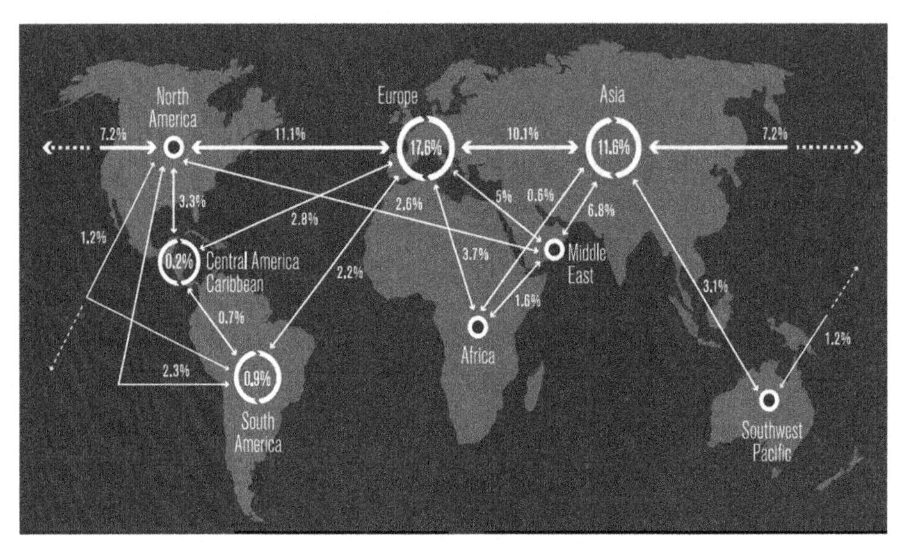

Fig. 3.5 The principal passenger traffic flows (in RPK) in 2015
Source: IATA WATS.

Table 3.12 **Ranking of airports per passenger traffic in 2016**

	City—Airport	MPAX	Δ% 16/15
1	Atlanta—Hartsfield	104.2	2.6
2	Beijing—Capital	94.4	5.0
3	Dubai—Int.	83.7	7.2
4	Los Angeles—Int.	80.9	8.0
5	Tokio—Haneda	80.1	6.4
6	Chicago—O'Hare	78.0	1.3
7	London—Heathrow	75.7	1.0
8	Hong Kong—Int.	70.7	2.9
9	Shanghai—Pudong	66.0	9.8
10	Paris—Charles de Gaulle	65.9	0.3
11	Dallas/F. Worth—Int.	65.7	0.2
12	Amsterdam—Schiphol	63.6	9.1
25	Madrid—Adolfo Suarez	50.4	7.7
43	Mexico—Benito Juarez	41.7	8.5

Source: Airline Business, May 2017.

countries, i.e., United States, Canada, the European Union, Japan, or Australia have experienced comparable traffic growths. Even Latin American airports that have shown rapid growth in the last years do not show currently comparable growth rates.

The list of airports is very different classifying them by cargo traffic (Table 3.14). The air cargo traffic volume is larger in Asia, where because of the long distances and

Table 3.13 **Ranking of Airports Per Passenger Traffic Growth in 2016**

	Airport (city)	Δ% 16/15	MPAX 2016
1	Kempegowda (Bangalore)	22.5	22.2
2	Don Mueang (Bangkok)	21.3	34.7
3	Indira Gandhi (Delhi)	21.0	55.6
4	Doha Int (Doha)	20.5	37.3
5	Incheon Int (Seoul)	17.2	57.8
6	Lukou Int (Nanjing)	16.7	22.4
7	La Guardia (N. York)	14.2	29.8
8	Huanghua (Changsa)	13.8	21.3
9	Jeju Int (Jeju)	13.2	29.7
10	Xianyang Int (Xian)	12.2	37.0

Source: Airline Business, May 2017.

Table 3.14 **Ranking of airports per cargo traffic in 2016**

	City—Airport	Tons (thousands)	Δ% 16/15
1	Hong Kong—Chep Lap Kok	4615	3.5
2	Memphis—Int.	4322	0.7
3	Shanghai—Pudong	3440	5.0
4	Seoul—Incheon	2714	4.6
5	Dubai—Int.	2592	3.5
6	Anchorage (including transits)	2542	−3.4
7	Louisville	2437	3.7
8	Tokio—Narita	2165	2.0
9	Paris—Charles de Gaulle	2135	2.1
10	Frankfurt—Rheim	2114	1.8
11	Taipei—Taoyuan	2097	3.7
12	Miami—Int.	2014	0.5

Source: Airline Business, May 2017.

the geographical obstacles the surface transport is less developed than in Europe. Six Asian airports are within the top twelve list. Two of the four North American airports in the list are the distribution centers of the two largest air cargo transport companies in the United States: Federal Express (Memphis) and United Parcel Service (Louisville). The case of Anchorage (Alaska) is particular because it has become a convenient intermediate stop for a big number of cargo aircraft flying from Asia to Europe or North America, while Miami is the main cargo center for Latin America.

Regarding economic results, the airport business provides better margins than other aviation industries, like the airlines or even the aircraft and engine manufacturers (Table 3.15) although the size of the airport companies or responsible organizations is small in terms of revenues compared to the other air transport industries.

Table 3.15 **Ranking of airport companies per revenues in 2015**

	Company	Revenues (MUSD)	Operating income (MUSD)	Net income (MUSD)
1	Heathrow Airport Holdings	4417	1401	1070
2	AENA	3885	1381	919
3	Aéroports de Paris	3216	868	474
4	Fraport aG	2866	574	328
5	Port Authority NY & NJ	2537	979	638
6	Hong Kong Int.	2344	1228	1080
7	NARITA Int. Airport Corp.	1819	361	202
8	Incheon Int.	1709	928	679
9	Japan Airport Terminal	1700	94	74
10	SCHIPHOL Group	1698	557	417
11	Airports Authority of India	1650	NA	387
12	Changi Airport Auth.	1561	NA	NA

Source: Airline Business, November 2016.

3.4 Air navigation

Civil aviation air navigation is regulated by the Chicago Convention. As a consequence of the recognition of the sovereignty of each State on their own air space, the elements for the compliance with Annex11 (where the structure of the air space and the air navigation procedures are established) must be equally provided by each State, covering the air space up to the limits with their neighbors. This statement does not preclude the constitution of multinational agreements to optimize the efficiency of the operation.

Particularly in Europe, the rapid growth of the international air transport and the introduction in 1957 of the first commercial transatlantic jets produced a certain congestion of the Upper Air Space (above 25,000 ft.), especially acute in the European zone of NATO, where the commercial and military traffic were mixed.

To address this issue, on the 13th December 1960 the International Convention EUROCONTROL was signed, creating the International Organization for Air Navigation Safety, generally known with the name of the Convention itself: EUROCONTROL. The agreement was ratified in March 1963 by six Member States: the Federal Republic of Germany, Belgium, France, Holland, United Kingdom, and Luxembourg. Ireland joint later in 1965. Another ten States (United States, Canada, Norway, Sweden, Denmark, Austria, Switzerland, Italy, Spain, and Portugal) adhered to the Organization between 1964 and 1967 as associates.

The foundational agreement was initially valid for twenty years (until 1980), renewable for five-year periods from that date. The objective was the complete unification of the member states services. However, the differences among the member states and the associates, every nation's own interests, and the high costs of EUROCONTROL limited the ambition of those objectives.

Although initially three Control Centers started to work, covering the Upper Air space of the seven founding states, two of them ultimately reverted to the national authorities, because France and the United Kingdom refused to transfer the control of their respective air spaces. Only Maastrich remained, controlling the Upper Air Space of the Benelux countries and the north of Germany, in the area of the Hamburg Flight Information Region (FIR).

In February 1981, a new conference held in Brussels agreed an extension of five years to the original agreement and a Protocol modifying such agreement. This Protocol was effective on 1st January 1986, opening the Convention to the adherence of new States and establishing as a long-term objective a Common Plan covering the complete Air Space (Upper and Lower) of the member states.

In order to face the increasing congestion of the European air space, in October 1990 EUROCONTROL announced the launch of the EASIE (Enhanced Air Traffic Management and Mode S Implementation in Europe) project. Among its objectives, to establish for 2005 a common and automatized control system in Europe, with a cost close to 1000 million ECUs.

The program addressed two main aspects:

- EATCHIP (European Air Traffic Control and Harmonization and Integration Program), started in 1990, to propose the technological standards and the future unified system architecture.
- APATSI (Airport/Air Traffic System Interface Strategy), dealing with the development of the at the time current system, trying to optimize the capacity of the existing infrastructures.

In June 1997 the EUROCONTROL agreement was reviewed, and its management capability enlarged with the inclusion of the gate-to-gate concept. In January 2000 ATM 2000+, a long-term project, was launched. ATM 2000+ aimed at covering the 2015 horizon, when the expected number of movements in the European air space was expected to double the number of movements in 2000. ATM 2000+ was part of the European Single Sky strategy, including among other programs, the satellite navigation system GALILEO. The strategy was to be developed by grouping gradually the existing national services in 9 Functional Blocks of Air Space (FABs), up to the complete fusion. At the end of 2010, the agreements to put in place the first three FABS were already signed: United Kingdom—Ireland, Denmark—Sweden, and the largest, Belgium—France—Germany—Luxembourg—Holland and Switzerland. They three cover more than 50% of the European flights.

In 2011, 39 States plus the European Union Commission were part of EUROCONTROL (including all the European Union Member States except Estonia) and participated in the collecting mechanism. This number of states is expected to increase as new agreements for the joint utilization of the existing means are being achieved.

EUROCONTROL is classified as an International Public Service Organization and their missions are:

- To ensure Air Navigation safety in their coverage zone
- Guarantee an equitable cost for all users of the system

The daily functioning of EUROCONTROL is managed by the Agency, reporting to the deciding organism, the Council, who is dependent on the political organism, the permanent Commission. The Council receives reports from four technical committees: coordination between civil and military controls, system security, comparative analysis of the performance of each service provider and common elements development (Maastricht/CEATS).

EUROCONTROL headquarters are in Haren, near Brussels, where their General Data Bank is. In March 1996, the Central Flow Management Unit (CFMU) entered into service in this location. The CFMU is a real-time simulation system of all movements in the European air space. It allows obtaining immediate solutions to any unforeseen contingency or the performance of planning studies on the effect of introducing changes in the services distribution.

The following so-called External Services depend also on EUROCONTROL:

- Control centers: Maastricht (Holland), operative from February 1972.
- Experimental center in Bretigny (France), active from January 1967.
- Air Navigation Institute in Luxembourg, inaugurated in October 1969.
- Central European Air Traffic Services (CEATS) in Budapest and Prague, created in 1999.

In 2010 EUROCONTROL had slightly over 3000 employees and their own operating costs were, approximately, 0.4% of the Member States total Air Navigation Aids costs. The overall budget of the organization is close to 700 million euros. The number of flights in the air space services by EUROCONTROL was 9.5 million, with an average delay of 2.27 minutes. The total amount of invoiced air navigation services was about 7400 million USD.

From the economic point of view, the EUROCONTROL base is the so-called Harmonized Regional system, created in November 1971, so that all the Member States share the same tariff system, corrected according to specific coefficients representing the cost differences among each State to provide the needed air navigation aids within their respective air spaces.

The way the navigation charges are established and applied to a particular flight is explained next.

The payment unit is called Service Unit, N.

$$N = d \times p$$

being

d a distance coefficient (equal to 1 for a 100-km distance)
p a weight coefficient, calculated according to the following formula:

$$p = \sqrt{\frac{\text{MTOW}(\text{ton})}{50}}$$

When a certain flight crosses the air space of any State, the charge it has to pay is:

$$r = t \times N$$

Where t is the coefficient corresponding to the air navigation costs of each State. The value of t is calculated automatically dividing the cost of the air navigation services of a given State among the traffic volume forecasted for a certain period of time. EUROCONTROL collects the money directly from airlines and pays back the corresponding amounts to the different States following the previously describe method.

Initially the t coefficient was calculated so that the States recovered just part of their investment. This percentage was 15% at the beginning and was gradually increasing, reaching 100% by 1983 in practically all Member States. This way, the air space users are paying for the totality of the air navigation aids cost.

This cost depends strongly on the economic conditions in each country (salaries, prices, interest rates) and the efficiency of their control centers. Generally speaking, it is considered that the EUROCONTROL prices are high, compared to other countries, while the service provided does not reach, especially with respect to congestion levels, the users' expectations. Despite air navigation charges in Europe being the most expensive in the World, the serious congestion problem in the air space suffered from 1988 has given a new impulse to the role of EUROCONTROL as the only solution to the harmonization of the European air traffic harmonization, fully supported by the European Union.

Price differences among the different States may be very important as seen in Table 3.16.

Table 3.16 **EUROCONTROL adjusted unit rates applicable to December 2017 flights**

Zone	Unit rate (EUR)
Portugal Santa Maria	10.06
Belg.—Luxembourg	67.53
Germany	69.43
Estonia	28.53
Finland	56.30
United Kingdom	72.79
Netherlands	66.33
Ireland	29.61
Denmark	60.58
Norway	44.81
Poland	43.96
Sweden	59.08
Latvia	27.53
Lithuania	44.49
Spain—Canary islands	58.43
Albania	51.09
Bulgaria	26.97
Cyprus	34.39
Croatia	46.00
Spain—Continent.	71.76

Continued

Table 3.16 EUROCONTROL adjusted unit rates applicable to December 2017 flights—cont'd

Zone	Unit rate (EUR)
France	67.07
Greece	30.02
Hungary	35.05
Italy	80.07
Slovenia	64.67
Czech Republic	44.53
Malta	18.86
Austria	72.78
Portugal Lisboa	40.19
Bosnia Herzegovina	43.32
Romania	32.28
Switzerland	97.80
Turkey	20.42
Moldova	62.33
FYROM	52.24
Serbia/Montenegro/KFOR	35.65
Slovak Republic	52.61
Armenia	36.95
Georgia	19.66

Aircraft design

4

4.1 Introduction

Aircraft design balances numerous and diverse criteria: payload (passengers and cargo), range, cruise altitude, cruise Mach, take-off field length, landing speed, cost, reliability, maintainability, etc. Fuel economy, together with emissions and noise solutions, must be compatible with all other requirements.

Modern commercial aircraft, at 75% load factor, burn typically between 2.5 and 3.0 L per RPK. This consumption is comparable or even better than that of a medium size car with 1.5 passengers. The commercial jet efficiency has improved more than 70% during the last 50 years, but it is just somewhat better than piston engine aircraft of that time.

A first and obvious line of action in order to reduce fuel consumption in the air transport industry is the development of more efficient aircraft from an energy point of view. The equation of the specific range (SR) is very useful to understand the fundamental aspects in the design of an aircraft contributing to reduce the fuel consumption:

$$\mathrm{SR} = \frac{a_0 M \dfrac{c_\mathrm{L}}{c_\mathrm{D}}}{\dfrac{\mathrm{SFC}}{\sqrt{\dfrac{T}{T_0}}} W}$$

where c_L is the aerodynamic lift coefficient; c_D is the aerodynamic drag coefficient; M is the Mach number; SFC is the specific fuel consumption; W is the aircraft weight; T is the static air temperature; T_0 is the static air temperature at sea level; and a_0 is the sound speed at sea level.

The SR equation shows clearly that the key parameters are aerodynamics ($c_\mathrm{L}/c_\mathrm{D}$), the aircraft weight ($W$), and the engine specific fuel consumption (SFC). From that equation, the improvement levels may come from these different elements:

- Aerodynamics (L/D)
- Weight (W)
- Propulsive systems (SFC)

These are the three large technological areas where progress has been achieved uninterruptedly from the 1980, and where research is still intensively pursued (Fig. 4.1).

A fourth are should be added to the previous three: the aircraft systems. They also contribute largely to the aircraft energetic efficiency because they allow a more precise navigation. However, the overall fuel burning saving potential is not the sum of the individual technologies. It depends upon the configuration of the aircraft and the integration of those in it.

Energy Efficiency in Air Transportation. https://doi.org/10.1016/B978-0-12-812581-6.00004-1

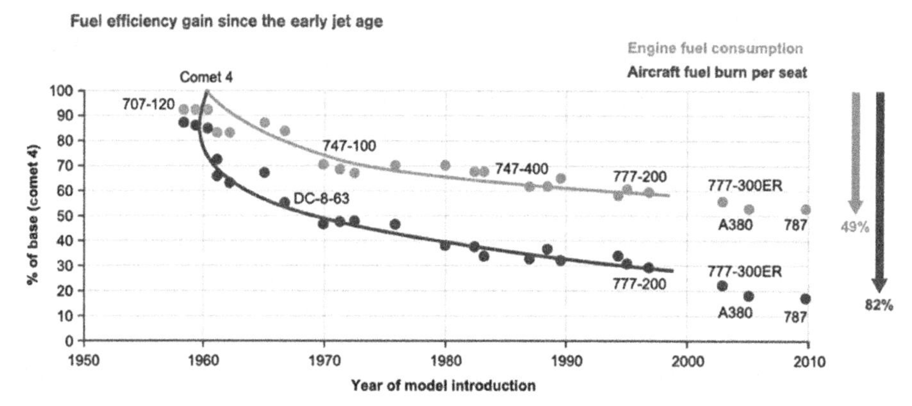

Fig. 4.1 Progress in fuel consumption (*CFM*).

New technological developments need time in order to be introduced in new aircraft models. This is why sometimes it is said that every new aircraft model enters into service with a technology level corresponding to approximately five years before. This consideration gives an idea of the importance of the development of new technologies and their industrialization, so they are available in the aircraft conceptual design phase.

These developments are terribly costly, and they are heavily supported by European (ACARE, Clean Sky) and American (CLEEN, PARTNER, ASCENT) research programs. These programs search for technologies to reduce fuel consumption by 50% in 2020 with respect to the technology in operation during 2000.

4.2 Technological areas of improvement

Probably the most relevant aerodynamic developments in order to reduce fuel consumption were supercritical profiles. This type of profiles, thicker than the conventional ones, had less curvature in the extrados, delaying the appearance of the shock wave in that zone. The consequence is the possibility to fly at higher Mach numbers, or at higher angles of attack, without increasing drag. In addition, in transonic regime, the shockwave formed in the extrados of supercritical profiles is weaker than in conventional profiles, causing a smaller wake and therefore a smaller aerodynamic drag. It is estimated that the utilization of this type of profiles in commercial transport aircraft produced an increase in energy efficiency around 10%.

Currently, improvements in aerodynamics are related to new numerical analysis developments and advances in the simulation of turbulence, like large eddy simulation (LES) or direct numerical simulation (DNS), together with the increase in the computing power of computers.

There are also new and more sophisticated tools for wind tunnel testing, such as laser velocimetry, particle injection velocimetry (PIV), cryogenic wind tunnels, and new visualization methods. All these advances are intended to achieve in the future the complete aircraft integrated design, i.e., the fuselage-wing-stabilizers assembly, in addition to new aerodynamic elements in the wing tip, such as the wingtips, winglets, or sharklets, or a better aerodynamic interaction between the wing and the engine nacelle.

For instance, the recent replacement of the wing fences by sharklets in the A320 model of Airbus results in a drag reduction that allows for an improved take-off and climb performance and up to 4% fuel burn savings and up to 1 EPNdB reduction on take-off compared with nonsharklet A320.

Other innovations for the future currently being investigated are (see Section 4.4 for more information):

- Smart wing technologies
  - Natural laminar flow
  - Hybrid laminar flow
- Morphing techniques for variable configuration aerodynamic surfaces

The second large technological area of efficiency improvement is propulsion. Historically, engine enhancements have been primarily responsible for the improvement of the energetic efficiency of air transport. Advances in the aerodynamic and thermodynamic design of the engine components, improvements in manufacturing processes, and the introduction of new materials have been the key elements to achieve those improvements.

The evolution of the energetic efficiency of different jet engines versus their bypass ratio is shown in Fig. 4.2.

The engine energetic efficiency in this case is defined as:

$$\eta = \text{Available Power} / \text{Consumed Fuel Energy} = TU_{\infty} / h_e \dot{m}_f$$

Where h_e is the fuel calorific power; $\dot{m}_f$ is the mass derivative with respect to time; T is the engine thrust; U_{∞} is the flight speed.

The graph displayed in Fig. 4.2 shows the great advance that the development of the turbofan engine had, with a continued increase in their bypass ratio up to values

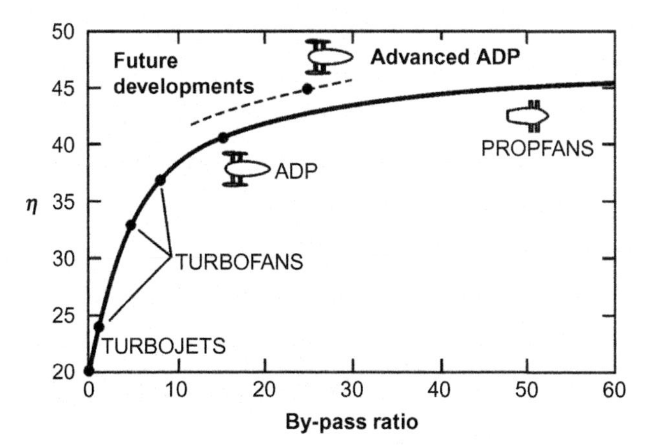

Fig. 4.2 Variation of the jet engines efficiency versus their bypass ratio.

between 5 and 8. The new developments in this type of engine are included in the graph, such as the advanced ducted propeller (ADP) and the propfan, also known as unducted fan (UDF). An unducted fan from GE was flight tested in 1990 in an MD80 (Fig. 4.3) with fuel savings up to 50% with respect to the JT8D-217D. It provided an acceptable economy starting at 90 US$/oil barrel. It showed though severe noise and vibration problems leading to certification uncertainty (blade failure case).

The ADP is actually a turbofan with a larger diameter fan allowing a bypass ratio in the range between 12 and 25, amply superior than that of current engines. ADP shows though a disadvantage compared to current engines, which is a larger weight and size. The Propfan is basically a turbopropeller, whose propeller, multiblade, has been designed to allow the aircraft fly at Mach numbers close to 0.8, without losing propulsive efficiency because of shock waves appearance.

Unfortunately, the situation in the technological evolution of the combustion process of jet engines implies an increase in the nitrogen oxides (NOx) emissions when the energetic efficiency improves, because this energetic efficiency is related to the decrease in carbon dioxide (CO_2) emissions. The increase in the engine's pressure ratio produces a beneficial effect on the fuel consumption because of the air cooling at the combustion chamber entrance, but it is prejudicial because due to the higher combustion temperature, more nitrogen oxides are emitted. Current research lines aim at allowing a better combustion process without increasing the air compression at the entrance of the combustion chamber in order not to increase the NOx emissions. New materials are being investigated to that effect, together with better technologies with regard to combustion processes in cooler air. An example of development with this aim is the multipoint lean premix (LP) combustor, tested in the CFM TECH56 TAPS Combustor showing 60% lower NOx than the current standard.

Fig. 4.3 Unducted fan from GE (*General Electric*).

Similar to the case of the fuselage-wing-stabilizers assembly, the improvement in the aerodynamics of the primary and secondary fluxes plays a very important role in the energetic efficiency increase. To this aim, the already-mentioned research in both numerical and experimental aerodynamics is being applied. Additionally, engine weight reduction by the introduction of new composite materials in its construction is also improving the efficiency. This new composite materials refer to carbon fiber materials utilized in the engine nacelles and fairings, but also to the development of metallic-matrix composites and ceramics to manufacture the engine components working under the most demanding conditions, i.e., blades, exhaust nozzle, and the fuel injectors or the combustion chamber in general.

An example of these new technological advances can be observed in the new Airbus A320 neo (new engine option), compared to the conventional A320 or as it is called now the A320 ceo (current engine option). The most significant differences come from the different aerodynamic devices in the wing tip (wing tip fences in the neo and sharklets in the neo) and especially the new engines with a better specific fuel consumption based mostly on their higher bypass ratio (12 in the case of the neo compared to 9 in the ceo). Based on these technological developments, according to Airbus, a fuel consumption reduction of 15% is achieved without the need to develop a brand new aircraft. The same approach has been followed by the other large aircraft manufacturer, Boeing, with the development of the B737 MAX from the existing B737.

In the ongoing large research programs (see Section 4.4), the concepts, which are being investigated under the name of Innovative Power Plant, are:

- Open rotor configuration
- Geared turbo fan
- Advanced turbo fan

Some of these innovations are being implemented in the latest products: the Pratt & Whitney 1000G (Gearfan) has already two certified versions and applications in the following aircraft: Bombardier C-Series, Mitsubishi MRJ, A320 neo, Irkut MS21, and Embraer 2. Other concepts that have been explored are the counterrotating fan, in conceptual phase, and tested in laboratory at component level, and the Aft fan, also in conceptual phase.

The third technological area of improvement concerns aircraft weight. Aircraft operational empty weight reduction seems to be unavoidably linked to the increase in the presence of composite material in their structure. From the 1980s, composite material utilization was extended in the aircraft construction (particularly CFRP or carbon-fiber-reinforced plastic, as a solid laminate or as a sandwich structure, but also although at a lesser extent GFRP or glass-fiber-reinforced plastic.

Initially some metallic elements of certain noncritical structures (those whose failure would not cause the loss of the aircraft), produced weight savings of up to 20% in those elements with respect to their equivalent ones manufactured in the traditional aluminum alloys.

Year after year this trend has continued increasing (Fig. 4.4), and successive models of, for instance, Airbus, have been increasing the percentage of primary structure manufactured in carbon fiber, reaching in the case of their most modern product, the

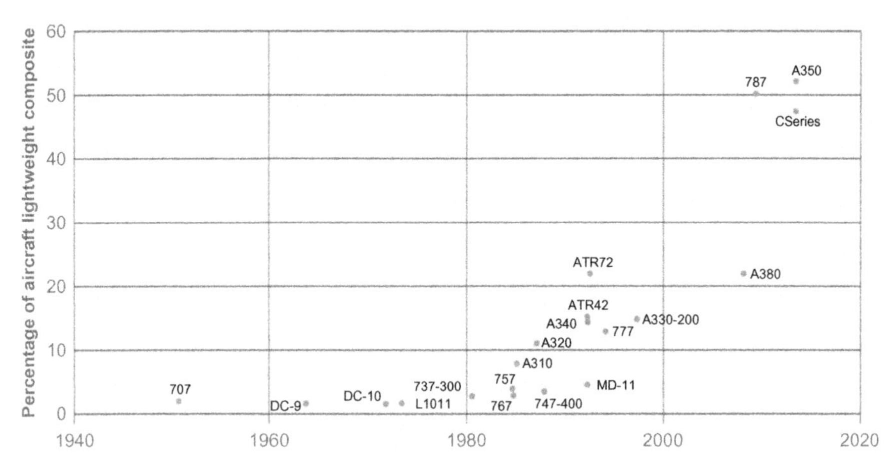

Fig. 4.4 Percentage of aircraft lightweight composite (*Airbus*).

airbus A350, up to a 53% of the aircraft structure weight manufactured in composite materials. Although Boeing was initially more reluctant to the utilization of composite materials in their aircraft, they are largely used in their latest models, and the new Boeing B787 that entered into service in 2011 has also about 50% of their structure weight in composite materials.

Future Innovative Structures include the following topics (see Section 4.4):

- New materials, composite, advanced alloys
- Self-healing, structural health monitoring
- Nanotechnologies

Finally, regarding aircraft systems, the introduction of digital electronics and computers on board must be mentioned. This is the so-called flight management system (FMS) and with it, fuel consumption reductions between 3% and 5% were achieved. Particularly important was the implementation in commercial transport aircraft of the fly-by-wire systems. In the fly-by-wire systems, the hydraulic systems are replaced by electric actuators, allowing the possibility to filter and improve the electric signal coming from the pilot controls. In this way, that signal can be adapted in real time depending on the actual flight conditions (possible increase in the atmospheric turbulence, aircraft center of mass displacement, etc.) responding thus with a better precision to the pilot's requests.

With the utilization of the fly-by-wire system, it is possible to control the stability and position of the aircraft center of gravity, the response under gusts and maneuvers, as well as the vibration modes and flutter (by the redistribution of mass, i.e., displacing part of the fuel volume). This enhanced control implies, in addition to an improvement of the flight operations safety, an important decrease in fuel consumption compared to previous hydraulic-systems-based technologies.

Fly-by-wire systems were first used in military aircraft and then in the supersonic aircraft Concorde. From 1985, they were also incorporated into commercial transport aircraft with the purpose of improving the aircraft controllability, when Airbus introduced the fly-by-wire system in the A320. Boeing introduced the system for the first time in the B777, already in the 1990.

From a different perspective, all the new electronic systems that allow a more precise navigation (Required Navigation Precision or RNP), play also a fundamental role in the improvement of the energetic efficiency of modern aircraft.

As discussed deeper in Section 4.4, the Systems for green operations research lines are:

- All electric aircraft
- Smart ground operations
- Energy management
- Fuel cells
- Satellite navigation allowing "free flight"

Some examples of how the aircraft manufacturers have incorporated technological advances in their new products are provided in this section, where public information released by the manufacturers has been used. According to Airbus published information, the new A350 XWB incorporates the following improvements, resulting in a 25% reduction in fuel burn:

- Airframe with advanced materials (53% composite, for instance a center fuselage CFRP panel 19-m long, 77m^2
- State-of-the art aerodynamics (M0.85 cruise speed)
- Simple, efficient systems
- Latest generation engines

Other results of the technological advances of the A350 XWB are the reduction in NOx emissions (35% below CAEP6 levels) and lower noise levels (up to 15 dB below ICAO Chapter 4 standard).

Another example, the A320neo burns 20% less fuel compared to the previous generation thanks basically to improved aerodynamics and new engines with higher bypass ratio, among other improvements. Additionally, the A320neo shows a reduction of NOx emissions 50% below CAEP6 levels and lower noise levels of up to 15 dB below ICAO Chapter 4 standard.

Boeing's 747-8 and 787 entered into service in October 2011. According to the company, the 747-8 Intercontinental carries more passengers for a longer distance while being cleaner, quieter, and more fuel efficient than any previous 747, while the 747-8 Freighter brings improved economic and environmental performance to the global cargo market. These new additions to the 747 family offer a 16% reduction in fuel usage and carbon emissions, and a 30% smaller noise footprint.

With a composite fuselage and wings, the 787 Dreamliner is 20% more fuel efficient than other previous airplanes of comparable size and has proved to be more environmentally progressive throughout the product life cycle. The airplane is manufactured using fewer hazardous materials, consumes less fuel, and produces fewer emissions. The Dreamliner also provides a 30% reduction in acoustic impact.

Boeing launched the 737 MAX in 2011, continuing the legacy by making the world's best-selling jetliner even more environmentally efficient. According to the manufacturer, airlines operating the 737 MAX will realize a 13% reduction in fuel consumption—with a corresponding reduction in CO_2 emissions—over the Next-Generation 737. Recent design updates, including the Advanced Technology winglet, will result in less drag and will further optimize the new engine variant's performance, especially at longer-range missions. The 737 MAX will also incorporate the latest in quiet engine technology, substantially reducing noise relative to current aircraft.

Despite all the efforts made by the aircraft manufacturers in the last decade to improve aircraft fuel efficiency working on all these mentioned technological areas, these efficiency gains are evolutionary. Revolutionary fuel consumption reductions would imply new aircraft configurations, different from the classical aircraft architecture of the last decades. These developments involve obviously a considerable risk for the aircraft manufacturers, and in a situation of sales increase and with a backlog of several thousands of aircraft to be delivered, investing heavily in new aircraft configuration does not seem to be the priority of Airbus or Boeing. Nonetheless, an important effort is put basically on government-funded research and development programs (see NASA Research in following pages) aiming at this new aircraft configurations. Probably the most advanced of these developments is the Boeing X-48C. The X-48C research aircraft flew for the 30th and final time on April 9, 2013, marking the successful completion of an eight-month flight-test program to explore and further validate the aerodynamic characteristics of the Blended Wing Body design concept (Fig. 4.5).

Another research line followed by aircraft manufacturers has to do with alternative propulsion systems, particularly with electric propulsion. Both Airbus and Boeing have carried out and still develop experiences aiming at increasing the TRL of promising technologies not yet mature. An example is the Boeing's Phantom Eye, an unmanned, liquid hydrogen-powered, high-altitude aircraft that according to the manufacturer can stay aloft for up to 4 days carrying a 450-lb payload. Phantom Eye began

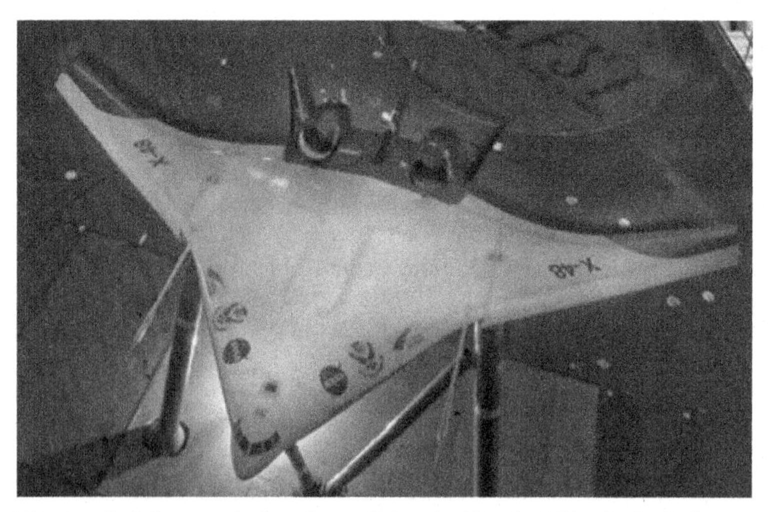

Fig. 4.5 Boeing X-48C research aircraft to validate the Blended Wing Body design concept (*NASA*).

flying demonstration flights in June 2012 and received "Experimental Aircraft Status" in United States by the National Aeronautics and Space Administration (NASA) and the Federal Aviation Agency (FAA).

Boeing also is researching the use of fuel cells—which convert hydrogen into energy and water with zero emission—for applications in aerospace and beyond. A remarkable experience in the utilization of fuel cells was the first crewed flight keeping a level cruise trajectory with fuel cells as the only power source performed by Boeing in April 2008.

Many of the new technological developments intend not only to improve the fuel efficiency but also reduce the environmental impact (see Chapter 9). For instance, SUGAR, which stands for Subsonic Ultra-Green Aircraft Research, is a contract that Boeing has with NASA to see what's out there in regard to technologies that might be viable for subsonic commercial aircraft to meet environmental requirements in 2030 to 2050. The SUGAR Freeze advanced commercial transport concept would use liquefied natural gas as a fuel. SUGAR Freeze is one of several notional concepts that a Boeing-led team is studying for NASA as part of the Subsonic Ultra-Green Aircraft Research (SUGAR) project.

4.3 Effect of fleet renewal in fuel efficiency

According to ICAO, its target is to achieve an annual 2% improvement in the world air transport system efficiency, measured in terms of fuel per RTK, in a period running until 2020 and continue at the same level during the next years, potentially until 2050.

The efficiency improvement is measured system wise and includes four types of improvements:

- Replacement of present operating fleet by new and more efficient models, consuming less fuel
- A more efficient use of the fleet transport capacity, increasing the average load factor of the commercial flights
- An air traffic management improvement, cutting in-flight delays and allowing airlines to operate in the most favorable track, wind, and temperature conditions
- A coordinated airport development in order to provide the requested capacity and minimize holdings and long taxi queues.

The most important of those elements is, no doubt, the fleet composition evolution. In general terms, new models are introduced into the market with a minimum of double-digit fuel consumption improvements with respect to the types to be replaced by them.

During the last 15 years, IATA has been recording the values of the revenue-ton-kilometer (RTK) of its affiliated companies, comparing them with the fuel consumption. The results show an average improvement in efficiency, measured in ton of fuel per RTK, around 1.5% yearly.

The ICAO Council, in its climate change mitigation program, set an aspirational target for the World Air Transport sector of 2.0% yearly improvement until 2020, with a possible stretch until 2030 if the national Action Plans prepared by contracting

States prove to be adequate. Also IATA considers possible an efficiency improvement of 1.5% per year in the future years and adopted this magnitude as a voluntary commitment.

The appearance of a number of new more efficient models in all size and range categories (A330neo, A350, B777-8 and 9, A320neo, B737MAX, C919, MS21, C-Series, and E2Jet) in the 2012–22 period and the promised improvements in the use of the air space (SESAR, NextGen) seem to support the credibility of that number.

The accelerated introduction of innovative aircraft types is reducing the average service life of commercial aircraft. According to Ascend, the average age of aircraft retired in 2015 was 25 years, coming down from 30 to 32 20 years ago.

As a characteristic example, Boeing fleet forecast for Europe in the 2016–35 period contemplates the delivery of 7570 new jets and the retirement of 4260 out of the 4610 in-service aircraft. According to that forecast, in the European 2035 commercial jet fleet, 95% will be new aircraft delivered in the 2016–35 period and only 5% were operative in 2015.

As described in Chapter 3, the commercial aircraft market is presently dominated by a small number of manufacturers that have been concentrated by acquisitions of or mergers with the majority of previously existing aircraft companies. Airbus (Europe) and Boeing (USA) are the two key players in the 120-650 seater area, with and output of 1400–1500 delivered new aircraft per year, while Bombardier (Canada), Embraer (Brazil), and Sukhoi (Russia) produce 160–180 regional jets yearly. Bombardier (Canada) and ATR (Europe) add other 140 turboprops in the 50–80 seat range.

In spite of this low number of manufacturers, the level of competition is strong and the technology has a continuous evolution producing aircraft with a higher energetic efficiency. Even with oil prices as low as today's 55 USD/Brent barrel, fuel stays as an important part of the airlines expenses, in the order of 19%–20% of total costs, and remains as a relevant marketing tool.

In the 2012–22 period, a completely new generation of commercial aircraft is replacing existing models. The new models that have entered or will be entering into service in that period are already in commercial utilization or in flight test, or within a launched program with firm orders. Therefore, their main features are known with reasonably good precision and it is possible to evaluate their improvements in terms of fuel efficiency. In addition, there are other programs in study, with longer-term temporal target. Their fuel efficiency is just a target and can only be used as indicative in the evolution of that characteristic.

According to ICAO statistics, the average real useful life of a modern airliner is around 25 years, measured by the age of aircraft at disposal. This is a lower figure than the one shown in a decade ago computation, being around 30 years. The difference indicates that commercial aircraft are arriving to their economic obsolescence much earlier than to their technical life limit (probably about 40 years), driven by intense airline competition and the cost of kerosene.

In order to ascertain the rate of aircraft replacement in the European aircraft fleets, an analysis has been made, taking data from Airbus Global Market Forecast 2016, Boeing Current Market Outlook 2016, and the latest ICAO and IATA statistics, covering the 2015–35 period.

The basic traffic growth figures are very similar in the four sources, and the global demand of new aircraft types is similar as well. Main differences between Airbus and Boeing come from the aircraft size mix to attend the same demand levels. Airbus forecasts fewer aircraft but with a higher average size, while Boeing calculates the delivery of a larger number of aircraft with a smaller average size. The differences in term of replacement rates are practically negligible.

The most relevant element in the composition of the commercial fleet operating in Europe is the European airlines fleet structure. The different information sources give very similar numbers of aircraft although divided in nonhomogeneous categories. In this analysis, the Boeing document Current Market Outlook 2016 has been used, because it contains only the fleet with registered flying activity during the year 2015 and does not account the number of parked aircraft (8%–10% of the total), with little chances to get back into service.

The different models are divided into six categories:

- Regional jets: single aisle up to 100 seats
- Narrow bodies: single aisle more than 100 seats
- Small wide bodies: twin aisle up to 300 seats in a three-class configuration
- Medium wide bodies: twin aisle between 300 and 400 seats in a three-class configuration
- Large wide bodies: twin aisle with more than 400 seats in a three-class configuration
- Turboprops are not included in this categorization, but their numbers have been added, using the Flight Global Ascend database (July 2016).

The production status of those models in operation or development at the end of 2015 is indicated in Table 4.1.

In the European airline operation, at the end of the year 2015, there were about 5400 aircraft in service in these categories. According to the fleet evolution forecast, that number will rise up to 7920 at the end of 2035, distributed in the way shown in Table 4.2.

There are two additional pieces of information to be added to the table for a better understanding:

- Although the number of very large aircraft decreases in the 2015–35 period, the total number of wide bodies doubles and their share in the number of aircraft in the fleet goes up from 21% in 2015 to 23.3% in 2035. This assumption is conservative, because it implies that airlines delay the introduction of larger-than-400-seat models until 2035. A faster introduction of new technology (A380 neo and its possible counterpart by Boeing) would improve the energetic efficiency.
- Regional jet number goes down because the trend to increase the size of aircraft in regional routes with models of 100 or more seats (CRJ 1000, CS-100, Embraer E190E2 and 195E2). The present 50-90 seat jets are replaced in many cases by turboprops, improving the fuel efficiency as well, but the total number of turboprop decreases due to the already-mentioned size increase effect.

As shown, 4910 aircraft (more than 90% of the European fleet operating at the end of 2015) will be retired in the 20 years period going up to 2035. In addition, traffic growth will require 3060 additional units, increasing 61.9% the total fleet number.

In order to calculate the effect of this fleet evolution on global fuel efficiency, it has been assumed that present aircraft are replaced by the new model with the

Table 4.1 **Production status at the end of 2015**

Aircraft category	Out of production	In production	In development
Turboprop	Antonov 24/32/38 Bae ATP, Jetstream 31/41 Beechcraft 1900C/D Bombardier Dash8-100/200/300 De Havilland Twin Otter Dornier Do 228 Embraer 120 Fairchild Metro Fokker F27/50 SAAB 340/2000	ATR 42/72 Bombardier Dash-8/400 De Havilland Viking	Leonardo TP90 Bombardier Dash-8neo
Regional jet	Avro RJ 70/85 Bae 146-100/200 Bombardier CRJ100/200/700 Dornier 328JET Embraer ERJ 145 family Fokker 70 and F28	Antonov 148/158 AVIC ARJ700 Bombardier CRJ900/1000 Embraer 170/175 Sukhoi Superjet	Embraer E2-175 Mitsubishi MRJ 70/90
Single aisle	Boeing B707/717/727/757/737 series 100–500, DC9, MD80-90 BAe146-300 BAeRJ100 Embraer 190/195 Fokker 100 Ilyushin Il-62 Tupolev Tu-134/154 Yakolev Yak-42	Airbus A318/319/320/321 Airbus A320neo Boeing B737-600/700/800/900 Bombardier CS100/300	Airbus A319/321neo AVIC ARJ-900 Boeing B737-MAX7/8/9 COMAC C919 Embraer E2-190/195 UAC MS-21
Small wide bodies	Airbus A300/310 Airbus A340/200/300 Boeing DC10 B767/200/300/400 Lockheed L-1011	Airbus A330/220/300/800/900 Airbus A350-900 Boeing B787-8/9 Ilyushin Il-96	COMAC C929
Medium wide bodies	Airbus A340/500/600 Boeing MD11 Ilyushin Il-86	Boeing B777	Boeing B787-10 Boeing B777-9/10 Airbus A350-1000 Airbus A350-2000 COMAC C939
Large wide bodies	Boeing B747/100–400	Boeing B747-8 Airbus A380/800	Airbus A380neo

Table 4.2 **European fleet evolution 2015–35**

Aircraft category	End of 2015	End of 2035	Delivered	Retired	% retired fleet
Turboprop	790	520	400	650	82.3
Regional jet	270	150	130	250	92.6
Single aisle	3370	5920	5880	3330	98.8
Small wide body	440	1140	960	260	59.1
Medium wide body	360	610	570	320	88.9
Large wide body	170	100	30	100	58.8
TOTAL	5400	8440	7970	4.910	90.9

closest features of size and range. With respect to the new models that will appear in the 2012–25 period, information has been gathered on their first commercial service date and their comparative energetic efficiency in comparison with the present models to be replaced by them. Information sources are the manufacturers' publications, specialized magazines, like *Flight International*, *Aviation Week and Space Technology*, or *Airline Business*, and customer airlines statements. The entry into service of aircraft in flight test or in project may suffer delays, but the global effect of aircraft replacement on fuel consumption does not change substantially.

In Table 4.3, new aircraft model appears classified by manufacturer, with their initial service date and a fuel consumption estimation with respect to the closest existing model.

The 2026–35 period is too far away for envisaging the performance of the programs to be launched by those dates, but the analysis assumes a similar rate of improvements than those in the 2015–25 period. (The only model announced is the Chinese 350-seater C939).

With the assumption that the replacement is done linearly (5% of the total number of aircraft is replaced each year), the energetic efficiency improves 1.1% per year. This leaves the other improvement factors (load factor, ATC, infrastructure) a 0.4% contribution to reach the 1.5% target, a magnitude lower than the 2% worldwide ICAO target but realistic, considering the maturity level of the European air transport system.

In a study published by the authors (Alonso et al., 2014; Aminzadeh et al., 2016), the evolution of the fleet composition of airlines operating in the EUROCONTROL airspace has been performed for the last decade, aiming at determining the improvement in the fuel efficiency due to the fleet renewal.

The evolution of the number of aircraft of each category is shown in Fig. 4.6, where the percentage of the number of flights operated by each aircraft type with respect to the total number of flights is shown from 2004 to 2013. The share of flights operated by wide bodies is relatively stable, being the same at the end of the period as at the beginning (12%), with a slight decrease during 2008 (10%) and the interval 2009–12 (11%). Probably the most relevant result is the sustained increase in the share of the single-aisle type, growing almost continuously from 62% in 2004 to 67% in 2013. This increase matches very well with the reduction in the share of regional aircraft, decreasing from a 26% share in 2004 to a 22% in 2013, where the decrease in the smaller

Table 4.3 **Fuel consumption estimation for new models in the 2015–25 period**

Aircraft type	Enter into service year	% Fuel consumption versus reference aircraft
ATR (Leonardo/Airbus) (Europe, 45–100 seat turboprop		
Leonardo TP90 (100 seater)	2025	1.00 ATR72/500
Airbus ***(Europe, 120–650 seat jets)***		
A319 neo	2018	0.90 A319
A320 neo	2016	0.89 A320
A321 neo	2018	0.88 A321
A330-800	2019	0.90 A330-200
A330-900	2018	0.88 A330-200
A350-900	2015	0.84 B767-400
A350-1000	2018	0.86 B767-400
A350-2000	2022	0.90 B767-400
A380 neo	2025	0.85 A380-800
Boeing (USA, 120–480 seat jets)		
B737 MAX7	2020	0.90 B737-700
B737 MAX8	2017	0.89 B737-800
B737 MAX 9	2018	0.88 B737-900
B787-9	2015	0.85 B767-300
B787-10	2018	083 B767-400
B777-8	2022	0.90 B777-300ER
B777-9	2020	0.93 B777-300ER
Bombardier (Canada, 70–150 seat jets, 80–90 seat turboprop)		
Dash-8 neo	2024	1.00 ATR72-500
CS-100	2016	0.90 E195
CS-300	2017	0.90 A319
COMAC-AVIC (China, 70–250 seat jets, 50–75 seat turboprops)		
MA70	2018	0.95 ATR72/500
ARJ700	2015	0.98 E175
ARJ900	2019	0.98 E190
C919	2018	0.90 A320
C929	2025	0.88 A330-200
C939	2030?	0.85 A330-300
Embraer (Brazil, 70–120 jets)		
E175E2	2020	0.90 E175
E190E2	2018	0.86 E190
E195E2	2019	0.90 E195
UAC-Irkut (Russia, 140–200 seat jets)		
MS21	2019	0.90 A320
C929 (cooperation with COMAC)	2025	0.88 A330-200
Mitsubishi (Japan, 70–90 seat jets)		
MRJ70	2021	0.88 E175
MRJ90	2020	0.86 E190

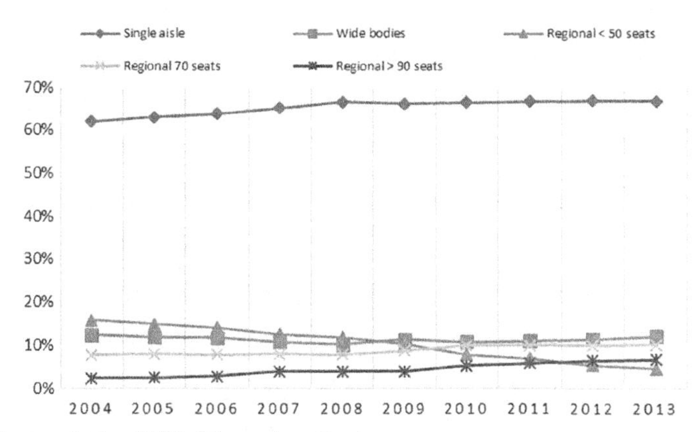

Fig. 4.6 Fleet evolution 2004–13 per aircraft category.

segment (<50 seats) from 16% to 5% does not compensate the increases in the 70-seat category (from 8% to 10% and in the more than 90-seat category (from 2% to 7%)).

The results show an overall increase in the average size of the aircraft flying these routes, where regional models are reducing their participation in the total number of the flights, while larger jets, both single aisle and wide body, are increasing theirs. These figures are consistent with the changes in the business model of the airline industry in Europe in the last decade, with the continued growth of low-cost carriers, typical users of single-aisle aircraft, and the difficulties of regional airlines caused at a great extent by the poor economy of small size jets in an expensive fuel environment and the competition of fast surface transportation modes. The improvement in the industry fuel efficiency shown previously can be partly explained by the evolution of the number of flights operated by aircraft of each category shown in Figs. 4.6–4.9, where the percentage of flights operated by each aircraft model with respect to the total of the corresponding category is plotted. Concerning single-aisle aircraft, the replacement of the classic B737 family by the newer B737 NG family is clearly

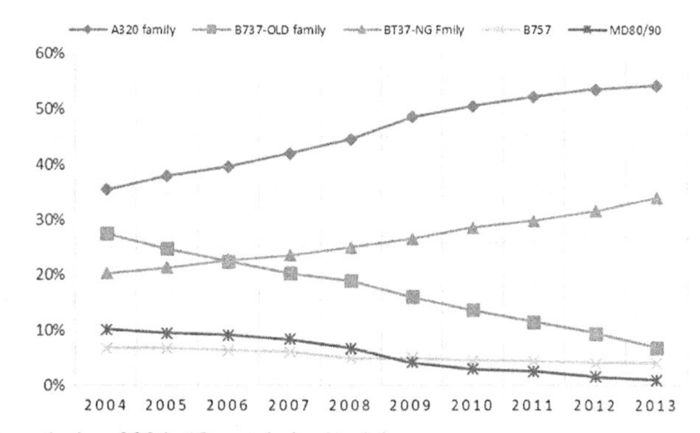

Fig. 4.7 Fleet evolution 2004–13, total single aisle.

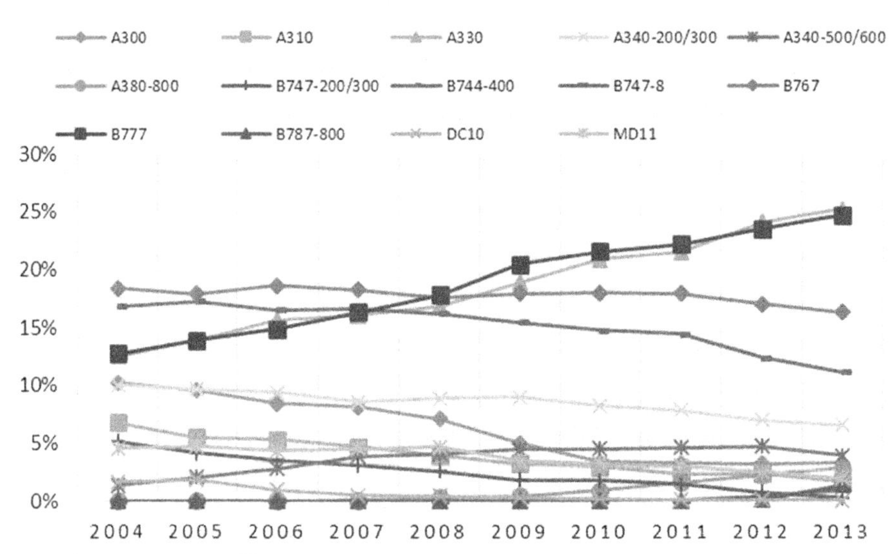

Fig. 4.8 Fleet evolution 2004–13. Wide bodies.

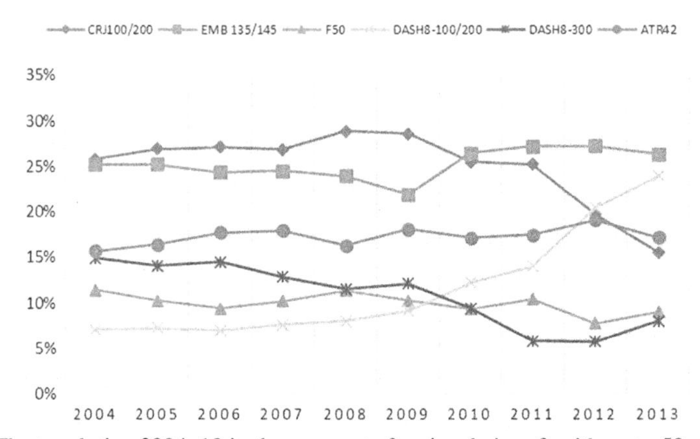

Fig. 4.9 Fleet evolution 2004–13 in the segment of regional aircraft with up to 50 seats (<50 seats).

appreciated in Fig. 4.7 as well as the reduction in the share of the B757 and the practical disappearance of the MD80/MD90 series, both replaced by a mix of B737NG and A320 families. All replacements are made with more energetic efficient models and collaborate to reduce CO_2 emissions per RTK.

Concerning wide bodies, as shown in Fig. 4.8, there is a continuous increase in the share of the two most modern long-range twin types, the A330 and the B777, each of them reaching 25% of the total number of flights of this category in 2013, starting in both cases also with a 13% in 2004. Only the A340-500/600 and the newest A380, B747-8, and B787 increase their share. There is a strong reduction in the relative number of flights of old models such as the A300/A310, the DC10/MD11 and the older versions of the B747 and A340-200/300, all of them out of production nowadays. The share of the B767 shows a slow decline from 18% to 16%. It can be concluded that along the full period 2004 to 2013 (and also between 2010 and 2013), there has been an important renovation in the commercial jet (single aisle and wide bodies) fleet operating in the EUROCONTROL airspace.

In the regional aircraft segment with up to 50 seats (<50 seats) (Fig. 4.9), there has been a strong increase in the share of turboprops, from 38% in 2004 to 49% in 2014, reflecting the need for efficiency improvement of regional airlines during the decade. It is particularly noticeable the increase in the share of the Dash8 family and the strong decrease of the share of the CRJ100/200, good example of jet-powered aircraft with 50 seats, affected by the high price of fuel and being partially replaced by more fuel-efficient turboprop models. The share of turboprops in regional aircraft segment with 70 seats, as seen in Fig. 4.10, has remained more stable varying around 70% for the whole period (adding the ATR42 and Dash8-400Q). From the jet side, there is a replacement of the CRJ700 and especially the older Fokker F70 by the Embraer E170.

Finally, in Fig. 4.11, the larger regional jets segment (>80 Seats) (8%), all of them have replaced the old (and far less efficient in terms of fuel consumption) Fokker F100, which had 100% of the share in terms of number of flights in 2004, and keeps just 11% in 2013. Summarizing, there has also been an improvement in fuel efficiency in the regional segment, coming from different sides:

- Replacement of jets by turboprops in the smaller segment (less than 50 seats).
- Replacement of old models (Fokker F70 and F100) by newer models (E170/190, CRJ900/1000) in the large regional aircraft segment.
- Overall, an increase in the aircraft size, with a net reduction in the number of flights operated by aircraft with less than 50 seats and an increase in the flights with aircraft with more than 90 seats.

The study (Alonso et al., 2014; Aminzadeh et al., 2016) shows that, due to these changes in the fleet composition, between 2010 and 2013 there is a reduction of 4.3%

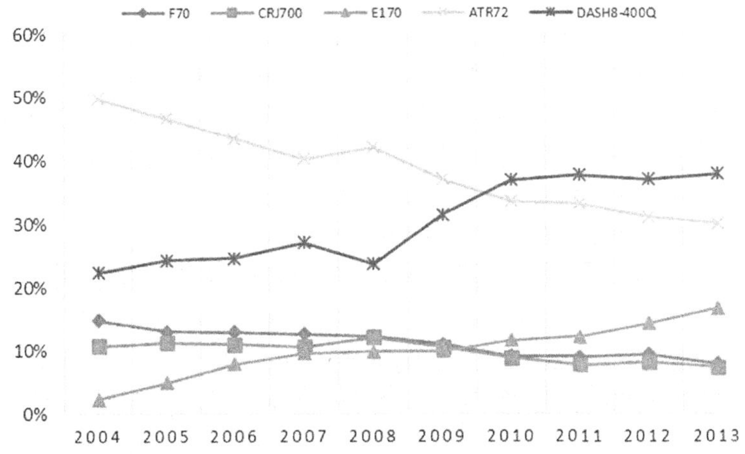

Fig. 4.10 Fleet evolution 2004–13 in the segment of regional aircraft with 70 seats.

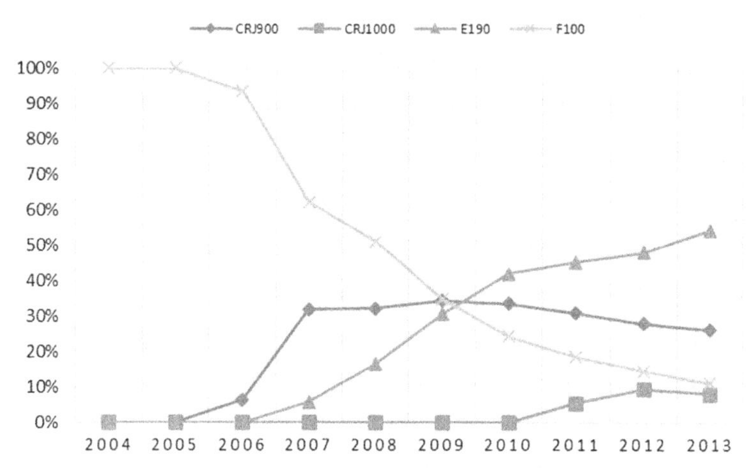

Fig. 4.11 Fleet evolution 2004–13 in the segment of regional aircraft with more than 80 seats (>80 seats).

in the CO_2 emissions, only partly explained by a modest decrease in the traffic (0.8% in terms of RTK). Overall, there is a 3.5% improvement in the efficiency parameter, measured in terms of emitted CO_2 per RTK (or burnt fuel per RTK), which corresponds to an average 1.2% yearly improvement.

The results of the study also show that longer distance routes (>2500 km) account for 73% of total CO_2 emissions, consistent with carrying 72% of the RTKs. Shorter distances (<500 km) are responsible for just 3.3% of the CO_2 emissions, but move 20% of the passengers. Most of the efficiency gain comes from the shorter-distance range below 1000 km (<1000 km) and especially from the long-distance range larger than (>2500 km). The methodology that has been followed to analyze the evolution of traffic and CO_2 emissions is also useful to derive the same type of information at airline level, and investigating for instance the differences in terms of energy efficiency among the various airline business models. The results are shown in Table 4.4, where the average efficiency is presented for a subset of network carriers, low-cost carriers, Regionals, and Inclusive Tour (IT) or charter carriers, a category still important in the intra-European traffic. According to this analysis, the average efficiency of the low-cost carriers represents 78% of the value for the network carriers, a very important difference that explains partly the cost advantage of this type of companies. ITs or charter carriers are even more efficient (60% of the efficiency of the network carriers).

Table 4.4 Airlines emissions and efficiency

Airline type	**Average efficiency (kg Co_2/ RTK)**	**Difference w.r.t. network carriers (= 100)**
Network carriers	1.06	100
Low-cost carriers	0.83	78
Regionals	1.84	1.73
ITs or charter	0.6	57

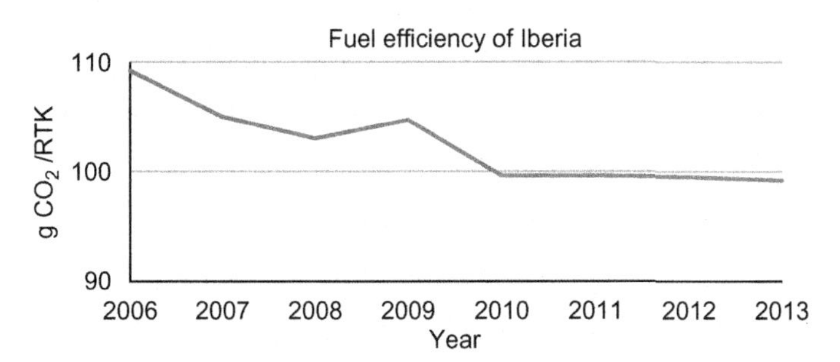

Fig. 4.12 Evolution of the fuel efficiency of Iberia (*Iberia*).

On the other hand, Regionals show the worst behavior, with efficiency 73% less than that of the network carriers. These results are consistent with the analysis of Miyoshi and Mason (2009) for the period 1997 to 2006. In order to investigate the reasons of those differences in the efficiency for each airline business model, the efficiency of the various aircraft types, depending on how they are utilized by each airline, has also been analyzed. A relevant aspect of this comparison is the higher efficiency (in terms of kg CO2/RTK) of the low-cost carriers compared with the traditional airlines when operating the same type of aircraft (Airbus A320 or B737NG family aircraft). Analyzing the dependency of the efficiency parameter with the average stage length, it is shown how the improvement in the efficiency parameter of the low-cost carriers compared to the network carriers flying the same type of aircraft is larger for longer values of the average stage length, but it is also the case even for smaller values of the average stage length, as a proof of a more modern fleet composition in the case of the low-cost carriers.

As a practical and actual example of the results of the fleet renewal policies that many airlines are carrying out throughout the World, the example of the Spanish carrier Iberia, founding member of the oneworld international alliance and part of the IAG holding is presented. According to information published by the company, Iberia has managed to reduce the CO_2 emissions thanks to the renewal in the long-range fleet and an improvement in the Load Factor of all their flights. The improvement in the fuel efficiency factor in the period 2006-2013 is 1.3%, consistent with the figure provided by IATA for the industry (Fig. 4.12). The company continues this process of fleet renewal: the long-haul fleet mix in 2014 was: 23% A330-300, 77% A340-300/600. By 2020 the mix will have become: 38% A340-300/300; 62% A330-200/300 and A350-900.

4.4 Large technology development programs in EU and USA

This Section provides a description of the large research and development programs being carried out both in the United States and Europe. These programs, funded in different ways by the respective governments, are shaping the evolution of technologies to be implemented in future generation commercial aircraft. Some of these programs are directly involved with technologies for aircraft design (for instance Clean Sky in Europe and CLEEN and NASA projects in the USA), while others are related to the air transportation system as a whole, also with very straightforward implications on aircraft certification. Examples are SESAR in Europe or NextGen in the USA.

The technology development programs respond to a strategy, which is ultimately defined by the governments, in very close relationship especially with the aircraft manufacturers but also with the rest of stakeholders. In 2001, the report of the Group of Personalities "European Aeronautics: A vision for 2020" pioneered an integrated vision of the European Air Transport System (ATS) for the next 20 years. It established, as its top-level objectives, the need to respond to society's needs and to secure European leadership in the aeronautics field. "Society's needs embrace the whole range of benefits that all citizens of Europe expect of the air transport industry now and in the future. These benefits are direct, as in the quality and price of travel, and indirect, as in the preservation of security and safety in a more global world. They encompass the personal needs of travelers and the collective needs of nontravelers who want to live in quiet, pollution-free neighborhoods."

The report also recommends the creation of the Advisory Council for Aeronautics Research in Europe (ACARE), the first European Technology Platform. European technology platforms (ETPs) were set up as industry-led stakeholder forums with the aim of defining medium- to long-term research and technological objectives and developing roadmaps to achieve them. Their aim is to contribute to increasing synergies between different research actors, ultimately enhancing European competitiveness. The Advisory Council for Aeronautics Research in Europe (ACARE) was launched in 2001 by then European Research Commissioner from the European Commission with 39 members, including representatives from EU Member States, EUROCONTROL, the European Commission, the European aeronautics industry, and air transport operators. The primary mission of ACARE is to define a Strategic Research Agenda (SRA).

ACARE has already produced a set of strategic research objectives (SRA1) in 2002 and a second updated edition (SRA-2) in 2004.

- SRA-1 is built around 5 Challenges for technology development. It has been used as a reference guide for a number of national and institutional bodies for establishing their research programs.
- SRA-2 describes 6 High-Level Target Concepts (HLTCs) and their associated technologies with respect to different socioeconomic scenario. Each HLTC stresses a particular aspect of the Air Transport System.

ACARE has been and remains a key guiding force in the planning of research under public, private, national, and EU programs. Its Strategic Research Agenda has also served as a major source of input in the formulation of the aeronautics work program of the Union's Research Framework Programmes (FPx). Major priorities remain the "greening" of transport (Fig. 4.13), strengthening competitiveness and efficiency, and responding to the increasing demand for mobility and higher safety standards.

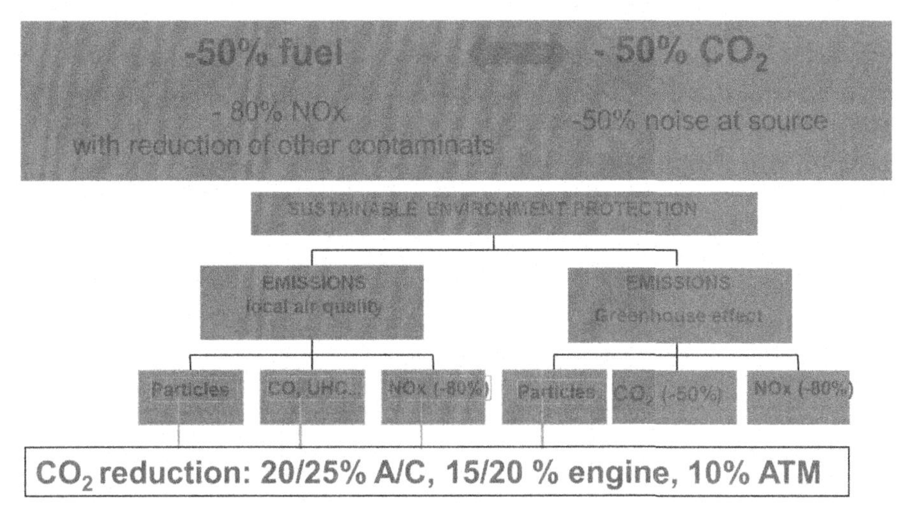

Fig. 4.13 ACARE program goals (*ACARE*).

The EU Framework Programme for Research and Innovation

The European Commission has been funding research projects under the so-called Research Framework Programmes. The last one was the Seventh Research Framework Programme (FP7) that has been replaced from 2014 with Horizon 2020.

Approved by EU member states on 3 December 2013, Horizon 2020 is Europe's biggest-ever research and innovation program with a budget of nearly €80 billion. It notes as one key societal challenge: smart, green, and integrated transport. This challenge aims to boost the competitiveness of the European transport industries and achieve a European transport system that is resource-efficient, climate-and-environmentally-friendly, safe, and seamless for the benefit of all citizens, the economy, and society.

The Transport Challenge is allocated a budget of €6339 million for the period 2014–20 and will contribute to four key objectives, each supported by specific activities.

Horizon 2020 will provide funding for a resource efficient transport that respects the environment by making aircraft, vehicles and vessels cleaner and quieter to minimize transport's systems' impact on climate and the environment, by developing smart equipment, infrastructures and services and by improving transport and mobility in urban areas.

Horizon 2020 also aims at a better mobility, less congestion, more safety and security with a substantial reduction of traffic congestion; with a substantial improvement in the mobility of people and freight; by developing new concepts of freight transport and logistics and by reducing accident rates, fatalities and casualties, and improving security.

Horizon 2020 supports a global leadership for the European transport industry by reinforcing the competitiveness and performance of European transport manufacturing industries and related services including logistic processes and retain areas of European leadership (e.g., such as aeronautics).

Horizon 2020 targets a socioeconomic and behavioral research and forward-looking activities for policy making. The aim is to support improved policy making, which is necessary to promote innovation and meet the challenges raised by transport and the societal needs related to it.

The topics for Aviation, within the Transport Challenge, are:

- 2016–17
  - Reducing energy consumption and environmental impact of aviation: 40 million €
  - Reducing aviation noise: 20 million €
  - Maintaining industrial leadership in aeronautics: 45 million €
  - Breakthrough innovation: 30 million €
  - Identification of gaps, barriers, and needs in the aviation research: 11 million €
- 2014–15
  - Competitiveness of European aviation through cost efficiency and innovation
  - Enhancing resource efficiency of aviation
  - Seamless and customer oriented air mobility
  - Coordinated research and innovation actions targeting the highest levels of safety for European aviation
  - Breakthrough innovation for European aviation
  - Improving skills and knowledge base in European aviation
  - Support to European aviation research and innovation policy
  - International cooperation in aeronautics with Japan
  - International cooperation in aeronautics with Canada
  - International cooperation in aeronautics with China

CLEAN SKY Joint Technology Initiative

The purpose of Clean Sky is to demonstrate and validate the technologies required to reach the ACARE targets. It covered initially the period 2008–15 stretched up to the end of 2015. The financing mechanism is 50% public-private, between the European Commission and the industry, up to 1600 M€. Clean Sky will contribute to meeting two of the ACARE HLTCs:

- Ultra Green Air Transport System, i.e., reducing the impact of air transport on the environment.
- Highly cost-efficient Air Transport System: Through innovative technological developments, Clean Sky will contribute to answering society's needs and securing European leadership in aeronautics.

Technologies allowing for the step change have to be developed, integrated, and validated to maximize the benefit of technology interaction and crossfertilization on the whole Air Transport System. Those are organized into six main themes, six Integrated Technology Demonstrators, ITDs, that cover the broad range of R&T work: aircraft—fixed wing (large and regional aircraft) and rotorcraft, engines, systems, and ecodesign concepts able to deliver more environment-friendly aircraft production and operations. A "technological evaluator"—a set of models to predict the local and global ecological impact of the technologies being integrated—will allow independent analysis of the projects as they unfold.

ITDs implement technologies, which, when brought together, will enable progress that will be truly groundbreaking rather than "incremental"—since reductions of around 30% are expected in terms of both carbon emissions and noise. These six themes are not just a simple "package" of programs that are independent of one another. On the contrary, they form a whole and the Joint Undertaking has to be structured in the most efficient way allowing both the full coverage of all areas of R&T work while ensuring a high degree of efficiency in the management of the technical activities.

Clean Sky is made up of six Integrated Technology Demonstrators:

- *SMART Fixed Wing Aircraft—SFWA*—will deliver active wing technologies and new aircraft configuration for breakthrough, new products.
- *Green Regional Aircraft—GRA*—will deliver low-weight aircraft using smart structures, as well as low external noise configurations and the integration of technology developed in other ITDs, such as engines, energy management, and new system architectures.
- *Green Rotorcraft—GRC*—will deliver innovative rotor blades and engine installation for noise reduction, lower airframe drag, integration of diesel engine technology, and advanced electrical systems for elimination of noxious hydraulic fluids and fuel consumption reduction.
- *Sustainable and Green Engines—SAGE*—will design and build five engine demonstrators to integrate technologies for low-noise and lightweight low-pressure systems, high efficiency, low NOx and low weight cores, and novel configurations such as open rotors and intercoolers.
- *Systems for Green Operations—SGO*—will focus on all-electrical aircraft equipment and systems architectures, thermal management, capabilities for "green" trajectories and mission, and improved ground operations to give any aircraft the capability to fully exploit the benefits of Single European Sky.
- *Eco-Design—ECO*—will focus on green design and production, withdrawal, and recycling of aircraft, by optimal use of raw materials and energies, thus improving the environmental impact of the whole products life cycle and accelerating compliance with the REACH directive.

A simulation network called the *Technology Evaluator* will assess the performance of the technologies thus developed.

The first phase of CLEAN SKY ended in 2013 and continues from 2014 under Horizon 2020 as CLEAN SKY 2. A continued JTI public-private partnership through Clean Sky 2 can deliver key outcomes:

- *Creating resource-efficient transport that respects the environment.* Building on FP7 achievements (particularly within Clean Sky), aeronautical research and innovation in Horizon 2020 must complete its role toward achieving the ACARE 2020 targets. Three-quarters will be achieved by the end of Clean Sky, with the JTI fully meeting its original goals. Beyond 2020, new efforts are needed to get within reach of the Flightpath 2050 targets of a 75% cut in CO_2, 90% in NOx, and a 65% noise reduction. Clean Sky 2 can facilitate the first steps on the way to these goals.
- *Ensuring safe and seamless mobility.* Aviation provides invaluable time-efficient mobility. New concepts can aid the air transport system in meeting evolving mobility needs of citizens: more efficient use of local airports, faster connections, and less congestion.
- *Building industrial leadership in Europe.* Clean Sky 2 will help maintain highly skilled jobs in Europe. Besides traditional rivals, Europe now faces strong competitors in “BRIC”

(Brazil, Russia, India, and China) countries. Clean Sky 2 will enable European industry to deliver the necessary innovations based on affordable and sustainable technologies. This will be supported by design tools and methods, program and supply-chain management, and certification processes that will shorten time-to-market, decrease nonrecurring costs, reduce risk, and create the global leadership essential for a sustainable industrial base.

These objectives will require both near-term solutions, which can be implemented in the next generation of aircraft, and breakthrough innovations to address the longer-term objectives—the air transport system from 2035 onward.

The initial Clean Sky Joint Technology Initiative has been followed by the so-called Clean Sky 2, with the purpose of developing flight test prototypes of the technologies developed in the initial Clean Sky (Clean Sky 1), for the period 2015–22, with possible extensions, depending on the achievements. The funding is still 50 public-private between the European Commission and the industry, up to 3600 M€.

The Clean Sky 2 programme consists of four elements:

- *Three Innovative Aircraft Demonstrator Platforms (IADPs)*, for Large Passenger Aircraft, Regional Aircraft and Fast Rotorcraft, developing and testing flying demonstrators at the full aircraft/vehicle level;
- *Three Integrated Technology Demonstrators (ITDs)*, looking at airframe, engines and systems, using demonstrators at major integrated system level;
- *Two Transverse Activities* (Small Air Transport, Eco-Design), integrating the knowledge of different ITDs and IADPs for specific applications and enabling synergies to be exploited between different platforms through shared projects and results;
- *The Technology Evaluator (TE)*, monitoring and assessing the environmental and societal impact of the technologies developed in the IADPs and ITDs.

The overall budget for Clean Sky 2 is nearly €4 billion. Of this, a total of €1.755 billion will come from the European Commission. The private sector industry leaders will contribute approximately €2.2 billion. This includes significant additional activities, which do not form part of the *Clean Sky 2* programme as described here, but which contribute to and support its objectives—enablers for the demonstrators or parallel research work necessary to develop an operational product in due time.

A maximum of 40% of the available EU funding (roughly €700 million) will be allocated to Leaders. At least 60% of EU funding—representing over €1 billion in EU funding—will be open to competition:

1. Core Partners compete via calls for Core Partners, becoming Clean Sky members once selected. The Core Partners will be eligible for 30% of EU funding, up to €540 million.
2. Partners compete through calls for proposals. The 30% of EU funding (€540 million) set aside for Partners will be awarded via participation in calls for proposals or, where relevant, via public calls for tender.

The structure of the program is presented in Fig. 4.14.

Brief overview of Clean Sky 2 programme components

- Large Passenger Aircraft IADP *(Leader: Airbus)*
- Regional Aircraft IADP *(Leader: Alenia Aermacchi)*
- Fast Rotorcraft IADP *(Leaders: Agusta Westland—Airbus Helicopters)*

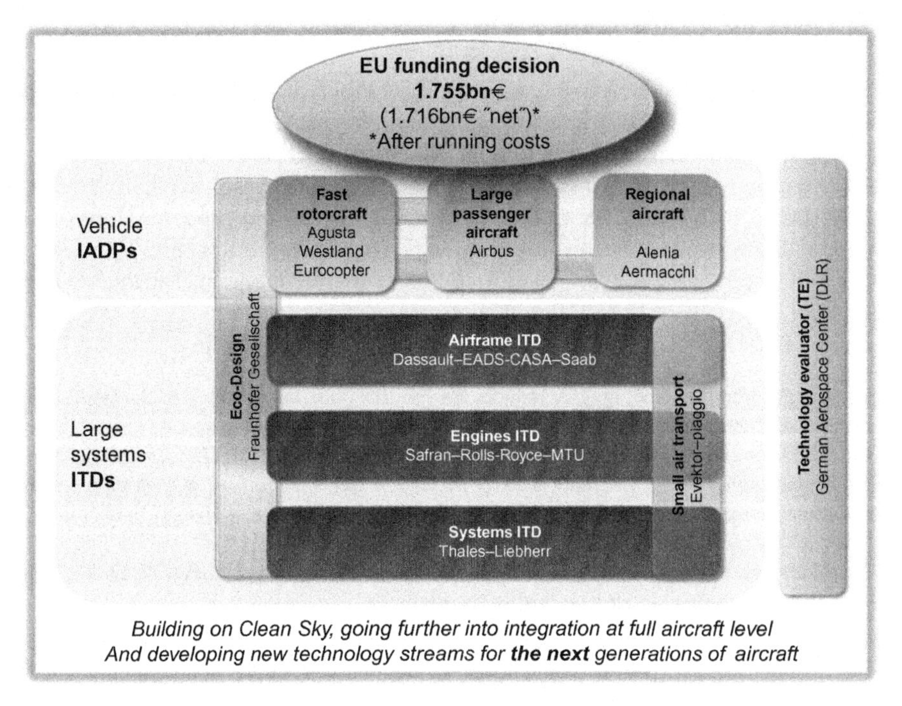

Fig. 4.14 The structure of the CLEAN SKY 2 programme (*Clean Sky*).

- Airframe ITD *(Leaders: Dassault Aviation—EADS-CASA—Saab)*
- Engines ITD *(Leaders: MTU Aero Engines—Rolls-Royce—Safran)*
- Systems ITD *(Leaders: Liebherr Aerospace—Thales)*
- Small Air Transport (SAT) Transverse Activity *(Leaders: Evektor—Piaggio Aero Industries)*
- Eco-DESIGN Transverse Activity *(Leader: Fraunhofer Gesellschaft)*
- Technology Evaluator *(Leader: German Aerospace Center DLR)*

CLEEN

The Continuous Lower Energy, Emissions, and Noise (CLEEN) Program is the FAA's principal environmental effort to accelerate the development of new aircraft and engine technologies and advance sustainable alternative jet fuels. Through the CLEEN Program, the FAA is a cost-sharing partner with industry. CLEEN projects develop technologies that will reduce noise, emissions, and fuel burn and enable the aviation industry to expedite integration of these technologies into current and future aircraft. CLEEN is a key element of the NextGen strategy to achieve environmental protection that allows for sustained aviation growth.

In 2010 the FAA initiated the first CLEEN Program, entering into five-year agreements with Boeing, General Electric (GE), Honeywell, Pratt & Whitney (P&W), and Rolls-Royce. These companies matched or exceeded the FAA funding in this cost-sharing program. Over the five-year period, the FAA invested a total of $125 million. With the funding match from the five companies, the total investment value exceeded $250 million.

Building upon the success of the initial CLEEN Program, in 2015 the FAA initiated a follow-on program, CLEEN II, which continues efforts to achieve the CLEEN goals and develop and demonstrate aircraft technology and alternative jet fuels. Under the CLEEN II program, FAA has awarded five-year agreements to Aurora Flight Sciences, Boeing, Delta Tech Ops/MDS Coating Technologies, General Electric, Honeywell, Pratt & Whitney, Rolls-Royce, and Rohr, Inc./UTC Aerospace Systems. These companies will match or exceed the awards in this cost-sharing program. The total federal investment is expected to be $100 million over five years.

The CLEEN Program goals include developing and demonstrating:

- Certifiable aircraft technology that reduces aircraft fuel burn, and/or supports the FAA's goal to achieve a net reduction in climate impact from aviation;
- Certifiable engine technology that reduces landing and takeoff cycle (LTO) nitrogen oxide (NOx) emissions below International Civil Aviation Organization (ICAO) Committee for Aviation Environmental Protection (CAEP) standards, and/or reduces absolute NOx production over the aircraft's mission;
- Certifiable aircraft technology that reduces noise levels, relative to the Stage 4 standard and/or reduces the noise contour area in absolute terms;
- "Drop-in" sustainable alternative jet fuels, including quantification of benefits—drop-in alternative fuels will require no modifications to aircraft or fuel supply infrastructure.

Quantitative goals for the fuel burn, emissions, and noise metrics under CLEEN I and CLEEN II are given in Table 4.5.

ASCENT

The Aviation Sustainability Center (ASCENT), also known as the Center of Excellence for Alternative Jet Fuels and Environment, is a cooperative aviation research organization funded by the FAA, NASA, the Department of Defense, Transport Canada, and the Environmental Protection Agency and coled by Washington State University and the Massachusetts Institute of Technology. ASCENT integrates 16 US research universities and over 60 private sector stakeholders with the purpose of reducing the environmental impact of aviation. ASCENT also works in partnership with international research programs, federal agencies, and national laboratories to create an all-inclusive research capability for dealing with the environmental impact obstacles the aviation industry faces. Approximate budget for the period 2014–16 is 30 M$.

Table 4.5 Quantitative goals for the fuel burn, emissions, and noise metrics under CLEEN I and CLEEN II

Goal area	CLEEN I goals	CLEEN II goals
Noise (cum below Stage 4)	−32 dB	−32 dB
LTO NO_x Emissions (below CAEP/6)	−60%	−75% (−70% re: CAEP/8)
Aircraft fuel burn	−33%	−40%

The Mission of ASCENT is:

- Focusing on meeting the environmental and energy goals of the Next Generation Air Transportation System, including reducing noise, improving air quality, reducing climate impacts, and energy efficiency.
- Exploring ways to produce sustainable aviation fuels at commercial scale, creating an industry with the potential for large-scale economic development and job creation.
- Discovering science-based solutions will benefit the aviation industry and improve the health and quality of life of those living and working around airports.

Research projects within ASCENT are divided into five categories: tools, operations, noise, emissions, and alternative fuels.

Research within the tools category involves researching current systems to understand the short- and long-term effects of new technologies. The ASCENT ream is working to develop tools to model and assess new and existing aircraft technology. Projects include:

- 010—Aircraft Technology Modeling and Assessment
- 011—Rapid Fleet-wide Environmental Assessment Capability
- 035—Airline Flight Data Examination to Improve Flight Performance Modeling
- 036—Parametric Uncertainty Assessment for AEDT 2b
- 037—CLEEN II Technology Modeling and Assessment
- 045—Takeoff/Climb Analysis to Support AEDT APM Development
- 046—Surface Analysis to Support AEDT APM Development

Research within the operations category involves improving aviation operations to reduce negative impacts on local communities, the environment, and the economy. The ASCENT team is working to develop efficient gate-to-gate aircraft operations, develop evaluation tools for aircraft performance, and explore new operations procedures. Projects include:

- 015—Cruise Altitude and Speed Optimization
- 016—Airport Surface Movement Optimization
- 023—Analytical Approach for Quantifying Noise from Advanced Operational Procedures

Research within the noise category involves researching noise pollution caused by the aviation industry. The ASCENT team is working to understand the impact of noise pollution on health, create tools for analyzing aircraft noise, understand how specific variables impact noise, and conduct outreach and education about aircraft noise reduction. Projects include:

- 003—Cardiovascular Disease and Aircraft Noise Exposure
- 004—Estimate of Noise Level Reduction
- 005—Noise Emission and Propagation Modeling
- 006—Rotorcraft Noise Abatement Operating Conditions Modeling
- 007—Civil, Supersonic Over Flight, Sonic Boom (Noise) Standards Development
- 008—Noise Outreach
- 017—Pilot Study on Aircraft Noise and Sleep Disturbance
- 023—Analytical Approach for Quantifying Noise from Advanced Operational Procedures
- 038—Rotorcraft Noise Abatement Procedures Development
- 040—Quantifying Uncertainties in Predicting Aircraft Noise in Real-world Situations
- 041—Identification of Noise Acceptance Onset for Noise Certification Standards of Supersonic Airplane

- 042—Acoustical Mode of Mach Cut-off
- 043—Noise Power Distance Re-Evaluation

Research within the emissions category focuses on reducing emissions from the aviation industry. The ASCENT team is working to analyze data and improve models to better understand the effect of airplane emissions, create and refine analysis techniques, and understand how policy changes could affect emissions. Projects include:

- 002—Ambient Conditions Corrections for Nonvolatile Particulate Matter (nvPM) Emissions Measurements
- 013—Microphysical Modeling & Analysis of ACCESS 2 Aviation Exhaust Observations
- 014—Analysis to Support the Development of an Aircraft CO2 Standard
- 018—Health Impacts Quantification for Aviation Air Quality Tools
- 019—Development of Aviation Air Quality Tools for Airport-Specific Impact Assessment: Air Quality Modeling
- 020—Development of NAS wide and Global Rapid Aviation Air Quality
- 021—Improving Climate Policy Analysis Tools
- 022—Evaluation of FAA Climate Tools
- 024—Emissions Data Analysis for CLEEN, ACCESS, and Other Recent Tests
- 039—Naphthalene Removal Assessment
- 048—Analysis to Support Development of an Engine nvPM Emissions Standards

Research within the alternative fuels category addresses the challenges associated with the creation and accessibility of alternative fuels. The ASCENT team is working to improve the feasibility of renewable fuels, understand how alternative fuels will affect emissions, air quality, and performance, and create standards for alternative fuel certification. Projects include:

- 001—Alternative Jet Fuel Supply Chain Analysis
- 021—Improving Climate Policy Analysis Tools
- 024—Emissions Data Analysis for CLEEN, ACCESS, and Other Recent Tests
- 025—National Jet Fuels Combustion Program—Area #1: Chemical Kinetics Combustion Experiments
- 026—National Jet Fuels Combustion Program—Area #2: Chemical Kinetics Model Development and Evaluation
- 027—National Jet Fuels Combustion Program—Area #3: Advanced Combustion Tests
- 028—National Jet Fuels Combustion Program—Area #4: Combustion Model Development and Evaluation
- 029—National Jet Fuels Combustion Program—Area #5: Atomization Tests and Models
- 030—National Jet Fuels Combustion Program—Area #6: Referee Swirl-Stabilized Combustor

Finally, the Evaluation/Support category includes the following projects:

- 031—Alternative Jet Fuels Test and Evaluation
- 032—Worldwide LCA of GHG Emissions from Petroleum Jet Fuel
- 033—Alternative Fuels Test Database Library
- 034—National Jet Fuels Combustion Program—Area #7: Overall Program Integration and Analysis

NASA AERONAUTICS RESEARCH MISSION DIRECTORATE PROGRAMS (ARMD)

In the USA, NASA has developed a new strategic vision for its aeronautics programs that are the culmination of a multiyear effort. Inputs were gathered from industry and other government agencies, including systems analyses of environmental and market trends, and the identification of societal megadrivers. The consensus was that NASA can best contribute to the nation's future societal and economic vitality by focusing aeronautics research in six thrust areas that are responsive to a growing demand for mobility, challenges to the sustainability of energy and the environment, and technology advances in information, communications, and automation. The six areas are:

1. Safe, efficient growth in global operations
2. Innovation in commercial supersonic aircraft
3. Ultraefficient commercial vehicles
4. Transition to low-carbon propulsion
5. Real-time system safety assurance
6. Assured autonomy for aviation transformation

To most effectively manage the research needed to address these six areas, NASA's Aeronautics Research Mission Directorate (ARMD) is structured into four mission programs:

- Advance Air Vehicles Program (AAVP)
- Airspace Operations and Safety Program (AOSP)
- Integrated Aviation Systems Program (IASP)
- Transformative Aeronautics Concepts Program (TACP)

The Strategic Implementation Plan sets forth the NASA Aeronautics Research Mission Directorate (ARMD) vision for aeronautical research aimed at the next 25 years and beyond. It encompasses a broad range of technologies to meet future needs of the aviation community, the nation, and the world for safe, efficient, flexible, and environmentally sustainable air transportation.

A key element of the plan is a set of Strategic Thrusts that are research areas guiding ARMD's response to global trends affecting aviation. This plan is a living document through which ARMD communicates with the aviation and research communities to elicit feedback. It will be updated as the conversation continues on achieving a vision that enables the current aviation system to best meet future demands and opportunities.

NASA, with aviation community input, has drafted a roadmap for each strategic thrust (listed earlier) to help guide research investments toward developing near-, mid-, and far-term solutions to aviation challenges. These roadmaps are the result of considerable thought within NASA informed by two National Research Council studies and many other interactions. The research and outcomes shown are intended to be community efforts and not research to be undertaken solely by NASA.

SESAR

As the technological pillar of Europe's ambitious Single European Sky (SES) initiative, SESAR is the mechanism, which coordinates and concentrates all EU research and development (R&D) activities in ATM, pooling together a wealth of experts to develop the new generation of ATM. Today, SESAR unites around 3,000 experts in Europe and beyond.

In 2007, the SESAR Joint Undertaking was set up in order to manage this large-scale and truly international public-private partnership. The total estimated cost of the development phase of SESAR is € 2.1 billion, to be divided equally between the European Union, EUROCONTROL, and the industry. Given the nature of the program and its scope, the Community contribution will come from Research and Trans-European Network funds.

It is in this context that the program aims to contribute to the SES high-level goals, by the end of STEP ONE of the European ATM Master Plan (Fig. 4.15).

These performance results are forecast to be achieved via R&D on improvements to ATM operations, including technical systems, procedures, human factors, and institutional changes as part of STEP ONE of the European ATM Master Plan. These targets represent approximately 50% of the SES High-Level Goals for STEP ONE.

In other words, the SESAR concept of operations will drive improvements to the procedures being used by all stakeholders, and in particular will start to modify responsibilities between technology, controllers, and flight crew.

NEXT GEN

The NextGen Office within FAA provides leadership in planning and developing the Next-Generation Air Transportation System (NextGen). As SESAR in Europe, NextGen is a transformative change in the management and operation of flights, which will reduce delays, save fuel, and lower carbon emissions. This comprehensive initiative integrates new and existing technologies, including satellite navigation and advanced digital communications. Airports and aircraft in the U.S. National Airspace System (NAS) will be connected to NextGen's advanced infrastructure and will continually share information in real time to improve air transportation's safety, speed, efficiency, and environmental impacts.

Fig. 4.15 SESAR objectives (*SESAR*).

The NextGen Office coordinates NextGen initiatives, programs, and policy development across the various FAA lines of business and staff offices. The office also works with other U.S. federal and state government agencies, the FAA's international counterparts, and members of the aviation community to ensure harmonization of NextGen policies and procedures.

NextGen Programs are:

- Automatic Dependent Surveillance-Broadcast (ADS-B) is FAA's satellite-based successor to radar. ADS-B makes use of GPS technology to determine and share precise aircraft location information, and streams additional flight information to the cockpits of properly equipped aircraft.
- Collaborative Air Traffic Management Technologies (CATMT) is a suite of enhancements to the decision-support and data-sharing tools used by air traffic management personnel. These enhancements will enable a more collaborative environment among controllers and operators, improving efficiency in the National Airspace System.
- Data Communications (Data Comm) will enable controllers to send digital instructions and clearances to pilots. Precise visual messages that appear on a cockpit display can interact with an aircraft's flight computer. Offering reduced opportunities for error, Data Comm will supplant voice communications as the primary means of communication between controllers and flight crews.
- National Airspace System Voice System (NVS) will supplant FAA's aging analog voice communication system with state-of-the-art digital technology. NVS will standardize the voice communication infrastructure among FAA facilities, and provide greater flexibility to the air traffic control system.
- NextGen Weather will help reduce weather impact by producing and delivering tailored aviation weather products via SWIM, helping controllers and operators develop reliable flight plans, make better decisions, and improve on-time performance. NextGen Weather is accomplished through collaboration between FAA, NOAA, and NASA.
- System-Wide Information Management (SWIM) is the network structure that will carry NextGen digital information. SWIM will enable cost-effective, real-time data exchange and sharing among users of the National Airspace System.

The United States and the European Union try to coordinate their actions in the area of air traffic management, trying to harmonize their developments. As a result, a Memorandum of Cooperation was signed in December 2014, the NextGen—SESAR Date: December 2014 Prepared by U.S.-EU MOC Annex 1—Coordination Committee State of Harmonisation Document.

Flight planning

5

5.1 Introduction

It is clear that no matter how efficient an aircraft may be in terms of fuel use, the actual fuel consumption in a specific flight will depend very much on how the vehicle is operated. A number of operational decisions are taken by airlines, but many of them are subordinated to the infrastructures they use, notably airports and air navigation services. The actual fuel consumption is therefore the result of the simultaneous action of the different stakeholders in the air transport industry.

In order to provide a systematic approach to this problem, the European Airlines Association (AEA) proposed a new policy in 2003, with the title of The Four Pillars, intended as an alternative to the proposal of imposing taxes on commercial aircraft emission levels, a regulatory initiative becoming popular among the European Governments at that time. This Four Pillars Policy (with a similar structure to the successful Balanced Approach Policy for noise reduction) suggests the urgent need of actions by all the stakeholders in four different areas:

- Research and development of better fuel consumption and emissions technology (airframe, engines, and systems).
- Infrastructure improvements (airports and air navigation services) in order to optimize flight operations.
- Optimization of operating procedures by the airlines.
- Introduction of market-based measures (MBM) if the previous three groups of actions were not enough to reach the mitigation goals

The role of aircraft and engine manufacturers regarding research and development of better technology has been already discussed in Chapter 3. Infrastructure improvements will be described in Chapter 8, while the most promising MBM are dealt with in Chapter 9.

Commercial flights are performed following a Flight Plan prepared by the operating airline. The main objective for the airline is to minimize the cost of the flight, fulfilling the applicable regulation and conforming to the airline's commercial and technical objectives. Fuel consumption reduction is a very important element to be considered when elaborating a Flight Plan, but it is not the only one, because minimum cost and punctuality are also relevant factors. The Flight Plan will be delivered to the airplane crew and to the ATM dependencies in the departing airport that will be providing the ATC services requested for that flight.

ICAO provides basic general rules for Flight Plans elaboration in the Annex 2 to the Chicago Convention. That document establishes that the Flight Plan is always convenient, but it is mandatory for all the flights in which:

- The aircraft receives Control services by the ATM
- In Instrumental Flight Regime (IFR), if the aircraft receives guidance by the ATM

Energy Efficiency in Air Transportation. https://doi.org/10.1016/B978-0-12-812581-6.00005-3

- To perform international flights
- In all cases when asked by the ATS

The content of the Flight Plan includes a number of administrative data to be communicated to the airport and ATM dependencies. The minimum amount of this information is:

- Date of flight in format YYMMDD (year, month, day; two digits each)
- Airplane identification (registration letters, tail number)
- Flight rules (VFR or IFR) and type of flight (scheduled, nonscheduled, general aviation, military, or others)
- Type of aircraft according to ICAO designation
- Wake turbulence category, for minimum separation purposes
- General equipment status (Communications/Navigation)
- Departure airport (four character location identification)
- Schedule block-off time, Coordinated Universal Time (UTC) in 24-h format HHMM (hour, minute; two digits each)
- Cruise speed as True airspeed
- Intended Flight level
- Route trajectory
- Destination airport (four character location identification)
- Total Estimated En route Time (EET)
- Alternate airports for that flight
- Total fuel endurance in format HHMM (two digits each)
- Total number of persons on board
- Emergency radio
- Survival equipment (jackets, number and capacity of the dinghies)
- Aircraft color and markings
- Any other relevant information
- Name and contact phone of the pilot in command

The main purpose of this group of data is providing to the ATM personnel information enough for monitoring the progress of the flight and provide adequate support in case of any unexpected incidence, including the possibility of an accident that requires the launching of a search-and-rescue operation. In addition, the Flight Plan will have other data, which will serve for the technical optimization of the flight, such as:

- The programmed trajectory for the flight, in the three spatial directions and with respect to the time
- The latest forecasted meteorological conditions
- The aircraft weight and balance status at take-off
- The fuel on-board the aircraft at any time during the flight
- The designated alternate airports for that particular flight

Annex 6 to the Chicago Convention establishes that the Flight Plan must be approved and signed by the pilot in command and by the flight operation officer or flight dispatcher. One copy must be left with the operator or designated agent, another one to the aerodrome head (or any other way of communicating to the departure centre). The Operations Manual of the operator shall include the contents and use of the Flight Plan, according to the procedures of the company.

ATM services may accept the Flight Plan and take note of its details or may interact with the operator and obtain a consensual modification of the original document. The case of EUROCONTROL illustrates this type of procedures. The multinational organization has a Network Manager Operations Centre (NMOC) receiving, processing, and distributing more than 35,000 flight plans a day in an area including more than 500 airports in 42 countries.

Every pilot on command submits a Flight Plan to the NMOC. The received data are checked against the airspace structure using the Integrated Initial Flight Plan Processing System (IFPS) model that shows the compatibility of that Flight Plan with the rest of requests received for the expected operating period. EUROCONTROL estimates that 9% of the Flight Plan produces inconsistencies to be resolved via manual interventions. Then, an alternative is suggested and, finally, a copy of the accepted Flight Plan is sent to all local air traffic control centers in Europe overflown by that particular flight.

The system has additional features, helping to correct possible similarities in airline call sign or to alert of the possible flight by an operator blacklisted from flying in European airspace. The continuity of the service is guaranteed by the parallel working of two centers, one in Paris and the other in Brussels, 24 h a day, 7 days a week.

The availability of high-speed and high-capacity computing systems allows multinational optimization of the traffic in a limited airspace with a large number of flights, generating wind optimal trajectories and optimizing any potential loss of separation between pairs of aircraft trajectories by small adjustments to departure times and small deviations from the previously wind optimized trajectories. In a majority of the cases, airlines do not perform a multipath optimization analysis for each flight. Short-range flights do not have many alternatives and attach to standardized trajectories unless something out of the norm appears. In the long range, optimization is more often used if alternative tracks are available. At this moment, there is a lot of research on the use of Big Data techniques for schedule airlines flight planning because the capability of storing and processing all main data of a huge amount of historical flights gives credibility to the idea of optimization by comparison, selecting the best historical Flight Plan with similar conditions of today's flight.

A good flight planning data warehouse helps improving risk management and allows Flight Dispatchers to search for the most efficient route, optimize the fuel loads, and minimize the risk of diversions. This task needs of appropriate tools including statistical information, having access to accurate weather and traffic information, and ensuring a close cooperation between pilots and flight dispatchers.

Schedule airlines have standard flight plans for their customary routes. Those "canned" flight plans are the base for the actual flight plan that is made by introducing actual flight values, like today's payload and meteorological forecast, to replace the standard plan values. This is a very fast and easy procedure for many short-medium range flights in which there is no alternate track to choose.

Although the basic procedure of Flight Planning is very much standard, there are some elements that offer a certain level of real-time choosing among several alternatives. Many of the practices and operational procedures described in this Chapter and in the next one are taken from the recommendations and the best practices suggested by ICAO and the International Air Transport Association (IATA) to their affiliated States and companies in different documents, like ICAO Flight Planning and Fuel Management Manual Doc 9976 or IATA's Flight Path to Environmental Excellence.

5.2 Trajectory optimization

The Flight Plan must optimize the trajectory of the aircraft, both in the vertical and horizontal planes. In the optimization process, the meteorological conditions and the indications from the ATM must be taken into account. In the horizontal plane, the trajectory tries to describe the shortest distance route (orthodromic). In the vertical plane, the airline may choose the minimum cost, the minimum fuel, or the target flight time criteria.

As a general rule, airlines try to apply the minimum cost criteria. The calculation is made by a simple model that compares the cost of additional fuel consumption for flying faster against the savings due to reduced flight time. This relationship is called Cost Index (CI) and it is discussed in point 5.3 of this chapter. If the price of time is small, minimum fuel burnt becomes the priority. A typical commercial aircraft jet reduces its fuel consumption with a higher altitude cruise, but different wind intensities may advice flying at a lower level in order to achieve minimum fuel burnt.

A third consideration is the flight time. Airline schedules publish Block times (time since the blocks are removed from the undercarriage to initiate the taxi out until the blocks are on at the arrival parking position). Punctuality is a quality greatly appreciated by passengers and by airline marketing people but, at the same time, delays may have high cost to accommodate passengers having lost their connections or even asking for economic compensations, more or less expensive depending on the regulatory conditions of the States in which the operation happened.

On the other hand, arriving early may not be a good business in congested airports, because the corresponding landing slot might not be ready and the aircraft has to be sent to a fuel-wasting holding until a free slot appears. Flight Plans use to calculate standard flight time before take-off and, later in-flight, a replanning can be done to compensate unforecasted events and keep the original arrival time.

The route selection depends on many different factors. It is normally the Flight Dispatcher who, with their choice of route, affects at a large extent the profitability of the flight. The effect is not the same for all routes; it depends very much on the length of the stage. For example, for a flight duration less than 2–3 h (short range flights), it is more difficult to save fuel in an individual flight, because there is no much time for it. However, many modern airlines have aircraft utilization of more than 10 h a day in this type of short- and medium-range flights. In this case, modest fuel savings in a single flight accumulate very rapidly for the whole fleet in a single day.

In the case of a high utilization series of short-medium range flights, with very demanding short turnarounds, the accumulation of small delays may endanger the whole rotation of the aircraft, because with short turnarounds there would not be time on ground to recover the delay. In this type of situations, the airlines change the flight plan target giving priority to the delay recovery and, therefore, fly at the highest technically recommended speed.

Generally, route selection is limited by the Air Traffic Control restrictions, both in the vertical and horizontal planes, and this is an important constraint for airlines when trying to save fuel. Situation is particularly serious in the case of high-density airspace and very often airlines are constricted to repeat the same trajectory day after day, without many chances to look for the optimum alternative. When planning a flight, all available options must be analyzed to ensure maximum fuel efficiency, because it could be the case in some occasions that several fixed routes are available between a given city pair. In those cases, the aircraft navigation capabilities need to be considered, because occasionally routes for a city pair may be designated for flight management system (FMS) aircraft only.

On the other hand, for long-range flights, modern fight planning systems can optimize for minimum cost using route specific Cost Index (CI), as described in the next Section, comparing the results of several flight tracks. Minimum Time Track (high CI) will often result in lower flight level, increasing therefore the fuel burn. En route step climbs must be planned where possible, to fly as close as possible to the optimal altitude. If unable to plan the most optimum altitude for fuel economy, it is usually better to opt for the best lateral profile and accept a slightly less-than-optimal altitude.

For greater accuracy in the fuel estimations, in the flight planned route the assumed take-off runway, the departure and arrival procedures and the landing runway should all be included as there may be nonnegligible differences in the flight length between use one runway or other, due to the diverse approach procedures. Noise abatement procedures are one of the main origins of those differences.

The amount of fuel loaded in an aircraft for a certain flight consists of the Trip Fuel (TF) plus the Reserve Fuel (RF). These amounts are determined in each case from the corresponding flight plan established for that flight. Under certain circumstances, it may be also necessary to upload more fuel because of fuel restrictions in some scales (although this situation results in a higher fuel consumption due to the increase in the aircraft total weight). This is called tinkering and it is explained in point 5.4.

The Trip Fuel (TF) is the fuel required to cover the distance between the origin and the destination according to the programmed flight without any deviation. The TF depends on:

- The aircraft weight at take-off
- The aircraft performance condition
- The distance to be flown
- The planned flight regime (flight speed and altitude)
- The meteorological conditions

The trip comprises all the phases included between the aircraft leaving the parking position at the origin airport and the stop at the parking position at the destination airport. The trip includes therefore not only the flight but the corresponding taxi in and taxi out at both airports and the landing and take-off runs. However, many airlines do not count taxi in fuel in the fuel trip, because it may be taken from the reserve fuel. The flight itself consists of the following segments: takeoff, initial climb, cruise climb, cruise, and descent, holding (potentially), approach and landing. In long-range flights, the initial cruise altitude is limited by the aircraft design and weight. After some time of cruising at that altitude, the burnt fuel reduces the total weight and the aircraft can go to a higher altitude. This practice is called stepped cruise. All these phases are shown in Fig. 5.1.

In addition to the Trip Fuel, it is necessary to upload a certain amount of additional fuel as reserves, in order to provide the crew with sufficient safety margin to face potential contingencies not foreseen when preparing the flight plan. A typical example could be a contingency at any moment during the flight that forces the aircraft pilot to land at an alternate airport. To guarantee the operational safety, the Reserve Fuel (RF) amount is determined following the regulations of aeronautical authorities (FAA in the United States, EASA in Europe), or international organizations (ICAO) and takes into account:

- The alternate airports for the different phases of the flight
- The holding over the alternate airport
- Any contingency out of the main Flight Plan, like errors in the meteorological forecast or traffic congestion

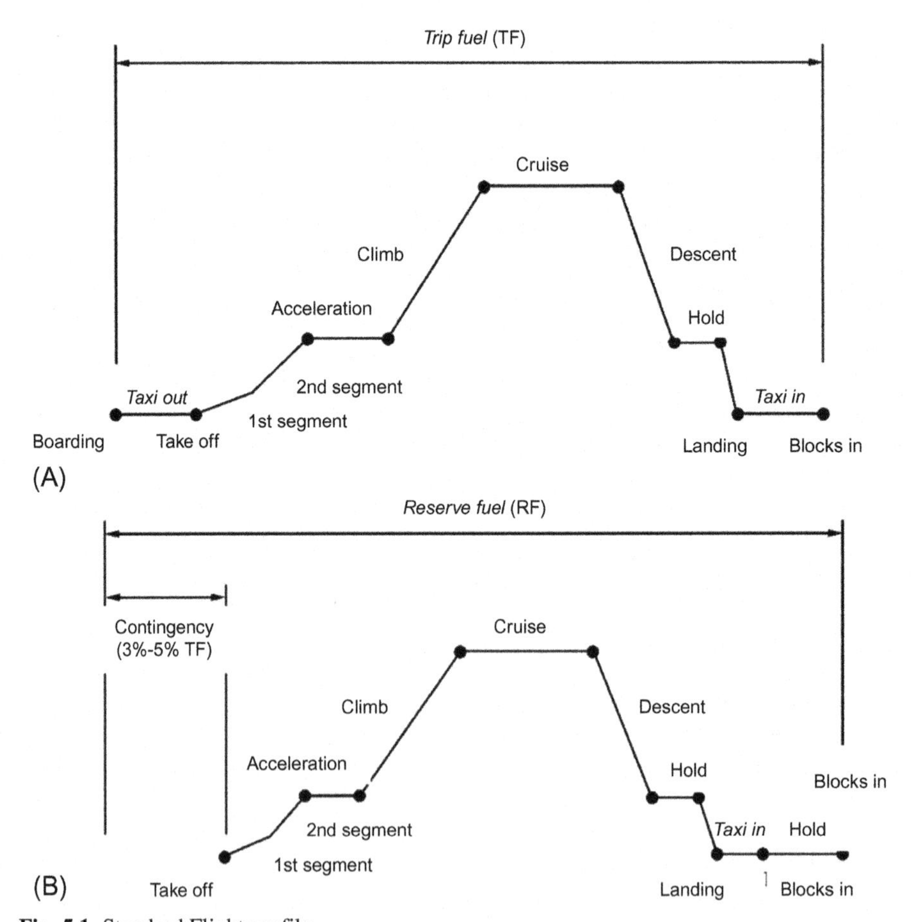

Fig. 5.1 Standard Flight profile.

The reserve fuel calculation is made on the assumption that the aircraft has to abort the landing at the destination airport in the very last minute and goes to the alternate airport at minimum consumption regime. Before landing there, a holding of 30 min at 5000 ft. over the airport covers the possible problems of the airport control to accept unexpected additional traffic. The flight profile corresponding to the trajectory from the aborted approximation at the destination airport runway to the alternate airport is shown in Fig. 5.2.

The amount of fuel loaded as reserves (RF) includes in addition to the fuel needed for the deviation to an alternate airport previously described, a small percentage, around 3%–5% of the trip fuel under normal conditions. This additional fuel covers for potential contingencies during the flight (weather conditions worse than anticipated, unexpected change in the cruise regime, etc.). Among the risk factor considerations are the possible deviations in the destination weather forecast at time of arrival or en route weather; turbulence at certain altitudes; lack of crew experience in that route; island or coastal destinations with worse wind conditions; and known ATC delays or frequent congestion at the arrival airport. More serious meteorological risks (blizzards, hurricanes, volcanic eruptions, earthquakes, and tsunamis) are more difficult to prevent, but there are other less dramatic like crosswinds, slippery runways, runway closure or special flight regimes like Extended Diversion Time Operations (EDTO is the present name, but it is sometimes refer as ETOPS or EROPS).

Basic twin engine commercial aircraft certification contemplates the possibility of one engine failure in flight. In this case, the pilot must go to an alternate airport that has to be no further than the distance the aircraft can go with a single engine running in 60 min. If the route goes over the ocean or desert areas, this regulation limits the possibilities of twin aircraft.

As the range of twins goes greater and the aircraft reliability improves, airlines asked to the CAAs to reconsider the 60-min limit. There is no general rule on the accepted time, but CAAs decide case by case, based on the demonstrated reliability

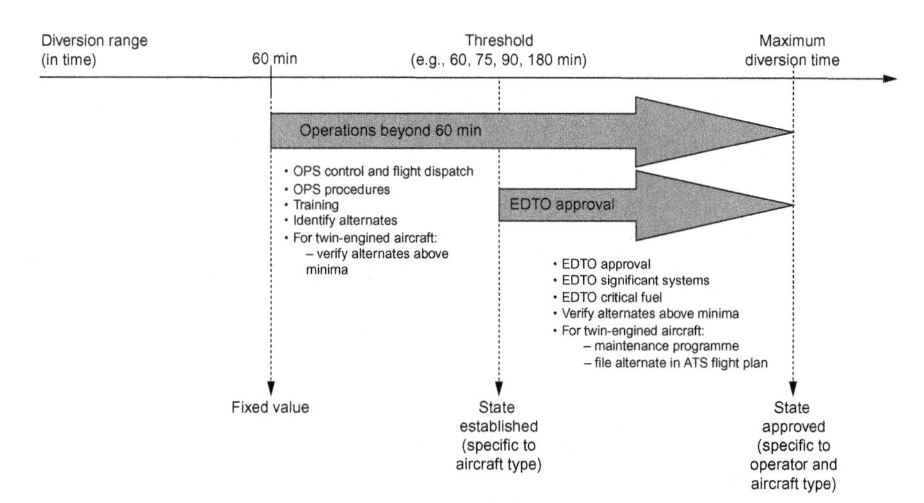

Fig. 5.2 EDTO scheme in ICAO Annex 6 (*ICAO: Annex 6 to the Chicago Convention*).

of the aircraft-engine combination. As the time to alternate increases, airlines can plan more direct flights with twins, as indicated in Fig. 5.3.

Present ICAO rules in Annex 6 Convention are merely indicative of the concepts and procedures and leave to the Type Certification CAA to approve a certain limit, based on the evidences provided by the aircraft manufacturer during the certification process. In general, other States CAA accept the original proposal, but they may introduce some precautionary measures, considering the experience of the operating airline in these types of operations. EDTO certification requires some additional aircraft equipment elements, special maintenance procedures, and a different Minimum Equipment List (MEL) than standard procedure. Then, the original EDTO time may be temporarily reduced until the operator gets enough experience.

A high time EDTO certification gives important benefit in terms of fuel efficiency as it allows the aircraft to fly shorter routes. Other important feature is providing more possibilities to the airlines to select the right capacity aircraft for each route without having in mind the number of engines. Fig. 5.4 shows a comparison of the limitations of EDTO 60, 90, 120, and 180 min. As seen, a 180-min EDTO offers an almost worldwide coverage, with the exceptions of mid-Pacific and the Antarctic Continent. Latest model aircraft are certified by the aircraft manufacturers with EDTO of up to 330 min, although 240 is the present standard for a long-range twin.

The EDTO issue is not yet finished as there might be potential new regulations depending on the utilization of remote aerodromes as on-route alternates. ICAO is discussing the features alternate airports should comply with, other than runway length and navigation aids equipment. Landing a large airliner in a remote, little-known, aerodrome without facilities to care for the passengers can be an undesirable solution.

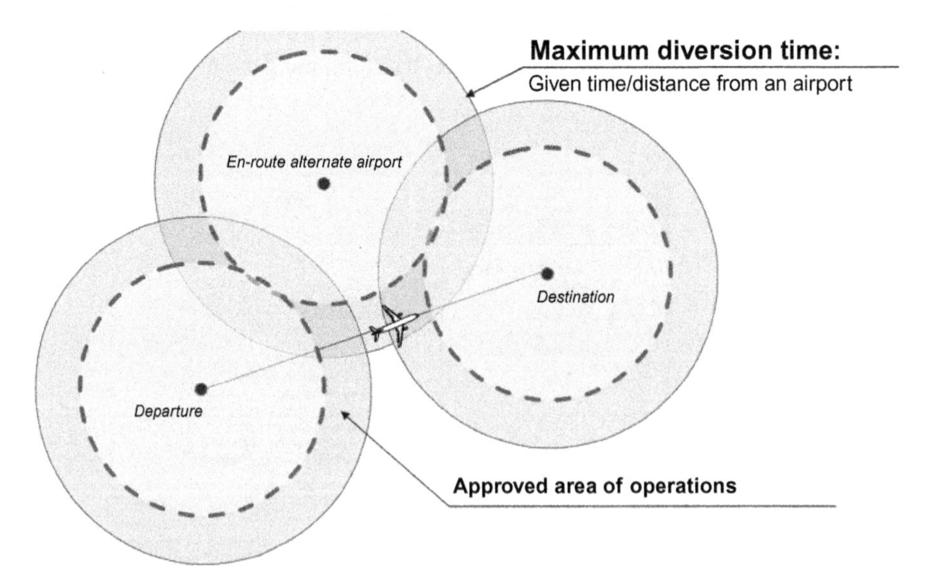

Fig. 5.3 EDTO application. Dotted line original certificated one engine distance. Continuous line new EDTO certified one-engine distance (*ICAO: EDTO Workshop*).

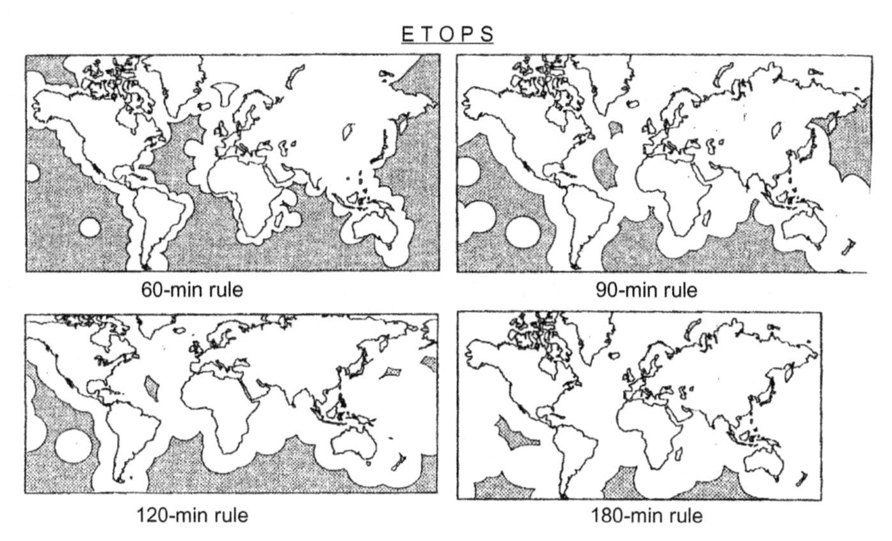

Fig. 5.4 Different ranges of EDTO coverage.

Finally, the aircraft pilot in charge of the flight has always the faculty to increase according to their best criterion the uploaded amount of fuel (on top of the trip fuel and reserve fuel). This is called discretionary fuel. As this is made to cover uncertain conditions, a key factor in reducing this additional fuel is communicating to pilots statistics on landing fuel and instilling confidence that the additional fuel is not necessary.

Currently many airlines use statistics to manage data such as taxi out and in times, historical flight time variations, contingency fuel and extra fuel boarding, and fuel consumption for individual aircraft. These data vary with the city pairs, the seasons, the day of the week, the time of the day, the flight crews, and the specific aircraft. An accurate tracking of planned to actual fuel loads can assist with fuel slip verification and fuel billing and accounting, but also help identifying where flight crews are adding fuel without coordinating with Flight Dispatch or Load Control. Additional fuel that is not included in the fight plan weights results in nonoptimized vertical and lateral flight profiles.

According to IATA, experience has demonstrated that without proper statistics, an average of 2 to 3 times the amount of discretionary fuel is carried compared to the amount determined from statistical information. A confidence factor covering 90% or 99% of these flights will demonstrate that in most cases, no additional fuel above regulated contingency fuel is required (4% or 5% of the planned flight fuel burn). Flight statistics help increasing the flight crew's confidence level of the flight planning system and will reduce their tendency of ad hoc fuel boarding. Many airlines use a 90% confidence factor during routine operation and 99% for IFR operation.

One of the key elements when planning a flight is the destination alternate selection. Actual diversions are required quite sporadically nowadays. Modern aircraft have auto land capabilities and operate at well-equipped airports with state-of-the-art approach facilities. Alternate weather requirements are very conservative and have not changed much in recent years despite the significant advances in aircraft navigation and landing systems, and improved weather reporting including satellite and radar imaging, as well as improved airport ground systems technology. Today diversions take place for a number of varied reasons, such as medical emergencies, aircraft maintenance, or weather conditions. As a consequence, most diversions occur to airports other than the alternate foreseen in the flight plan.

The destination alternates are chosen using one or more of these criteria:

- The closest to the destination airport to minimize the reserve fuel.
- One airport close to the destination airport but not having the same meteorology. If one of the reasons for going to the alternate is the weather over destination, it is necessary to minimize the probability of having the same bad weather in the alternate.
- One airport with the adequate ground equipment to serve the type of aircraft being deviated there. Some small airports may not have wide-body compatible ground equipment.
- One airport with good communication by alternative transportation means with the destination city in order to relocate passengers to their true destination as fast as possible. Particularly important in short-range flights.
- One airport with a wide range of connections for assigned new flights to the passenger not having destination airport as the final point of their trip, but just as a connecting point. This is important for long-range flights.

Airline Operation Manuals have two or three alternate airports in each route, classify by preference ranking, having considered all these factors. Unless something special is taken into account, the Flight Plan is made with the number one in the list.

5.3 Flight cost optimization

As a general principle, airlines try to obtain the maximum operating margin from their flight program. Revenue Management Systems and Flight Occupation Control are tools to obtain the maximum revenue. On the other hand, the Flight Plan must minimize the operating cost.

Basic airlines flight plans follow the minimum cost criterion. The Flight Planning system on ground and FMS on board can use a CI to calculate the optimum mix of different variable cost elements, mainly fuel consumption and flight time to minimize the trip cost. The minimum cost of a flight corresponds to a balance between the fuel cost and cost of time. The flight procedures are always limited by the safety margin conservation and the strict adherence to the ATM instructions. Cost-Index-based flight plans are available also for non-Flight Management Computer systems equipped aircraft, using paper printed tables.

The Cost Index represents the ratio between the cost of 1 h of operation and the cost of one unit of fuel (in weight). The two large commercial aircraft manufacturers use different units, but the concept is the same:

- Metric (Airbus): CI Model: Flight Cost ($/min) /Fuel Cost ($/kg)
- English (Boeing): CI Model: Flight Cost ($/h) /Fuel cost ($/100 lb)

The fuel variation with the speed depends on the aircraft technical features and the meteorological conditions and can be easily calculated using performance data. The fuel price may be slightly more difficult to calculate, when the price in the departure airport is different than the price in the airport of arrival. More fuel consumption may lead to more uplifted fuel in origin or more refueling at the arrival before the next flight. A general practice is using the first one for the economic optimization.

The variation of operating cost with time is more difficult to calculate, because the possible variations are marginal compared with the magnitude of total cost. The three more used elements to be taken into account are maintenance cost, crew cost, and delay and cancellation cost.

Maintenance cost variation depends on how much of the maintenance work is flight time dependent. In scheduled maintenance, there are some tasks performed when the aircraft gets a number of flights (undercarriage revisions, for instance) and others that are calendar based, like overhauls. These are not included in the CI. The third category is actions made on flight hour basis (most systems are that way). If the flight is longer, the cost of this group increases.

Crew cost variation with time depends on the airline type of contract with its personnel. If crew payment does not depend of the flight time, because crews have a certain amount of flight time included in their base salary, the change is nil. Many airlines have a fixed part of the salary and some extra for additional flight hours. Other airlines may rent flight hours from subcontractors and pay 100% of the cost as variable. Finally, in some companies, each route has a standard time to be computed for crew salary and small time differences have no economic repercussion. This is made to avoid problems when a route is operated by two different aircraft types with different speeds and it is convenient to pay the same to the crews. A possible additional factor to be considered is the consequences of flight duration in the regulated activity limits of the crew, although this is a factor not very common to have any influence in the CI.

The third and more difficult element to introduce in the CI are the costs caused by delays and cancellations (loss of connections, indemnities to be paid to the passengers, aircraft and crew turnaround costs, cost for airport staff) that may also depend on the flight time. There is a very wide range of cases—going from slight delays without having any economic consequences to cases in which the destination airport closes due to the end of the opening period or a night noise curfew and the flight has to go to an alternate airport. Between these two extreme cases, most normal events are delays with passengers losing their connections and having to rebook in other flights and/or having the right of receiving an economic compensation, depending on the type of flight and the geographical area. Airlines treat those events in a statistical manner, applying average time delay cost, depending on their experience. As those data are not published it is almost impossible give a general rule, but a reasonable approach may be to assign zero value up to 15-min delay and increase very much the value with more than 3-h delay.

A schematic idea of the differences between flying minimum fuel and minimum cost can be seen in Fig. 5.5. Flying faster than the Maximum Range Cruise (MRC point) regime increases the fuel consumption and reduces the flight duration. The distance between the minimum cost and the minimum fuel increases as the value of time goes up. The standard airline regime uses to be Long-Range Cruise (LRC), about 1% faster than MRC and the minimum cost point is not far away from it.

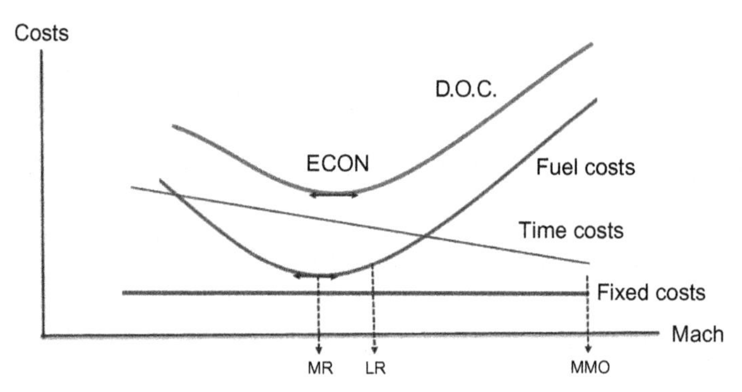

Fig. 5.5 Scheme of Cost Index basis.

The Cost Index can be generic for all the operations of the airline (common in short-medium range airlines flying a single aircraft type), or determined specifically for aircraft type, each group of routes or specific flight. As changing it very often can be difficult for the ground personnel, it can be modified by the crew or a generic one can be used by default. This tool also allows to check from the cockpit the details of the Flight Plan elaborated by the flight dispatcher.

Airlines with a low time cost structure and high fuel prices would normally operate at low CI and consequently lower speeds and vice versa. Since the price of fuel may be different at every airport, some airlines adjust the CI to be route specific. Therefore, an efficient flight planning system should have the full range of CI planning capability with appropriate vertical and lateral optimization. The vertical and lateral optimization should vary with CI as the winds and temperatures will vary with different altitudes. The vertical optimization should look both up and down as winds and temperature vary greatly at different altitudes.

By using the Cost Index, the crew can make the FMS optimize the operation for minimum fuel consumption, time or costs. Once the CI has been defined, the FMS control the flight parameters in all the phases (climb, cruise, and descend speeds, etc.), taking into account the actual aircraft and weather situation at any time. It is possible for instance in a headwind situation to make up flight time by accelerating and descending to minimize the effect of the head wind: in the case of an A330 in a 200-nm cruise segment at FL 350, accelerating from 0.80 M to 0.84 M can save 12 min in zero wind or 22 min with 120-kt headwind. However, in a tailwind situation, it is literally impossible to make up flight time by accelerating as descending will take the flight out of the wind.

A small value of the CI (close to 0) indicates a flight plan where the fuel cost is considered much more important than the flight time. It is typically an MRC regime operation. A large value of the CI (depending on the manufacturer, the CI is graded from 0 to 99 or to 999), on the other hand, gives more importance to the flight time than to the fuel consumption. It is therefore an operation minimizing the flight time and maximizing speed.

Cost Index system enables a unique flight path optimization solution, which can save money by achieving more efficient flight operations. The CI will enable an airline to achieve:

- Minimum cost operations (Normal Ops) by optimizing the fuel/block time tradeoff.
- Absolute maximum range from the aircraft for actual flight conditions (Minimum Burn Ops)
- Absolute minimum flight time for actual flight conditions (Minimum Time Ops)
- Dynamic flight regime actualization as the conditions in flight starts to deviate from the original flight plan
- Adjustments for optimizing cost, while fixing an exact arrival time

During preflight, the FMS should be used to crosscheck the flight plan time and fuel data. Many pilots have different methods of performing fuel checks, but many of these methods have limitations. Some crews use an average fuel burn figure per hour based on the flight time, for instance. Fuel performance charts have many limitations: specific airframe and engine in service deterioration, no adjustment for CI, no winds and temperatures at specific waypoints, last-minute Zero Fuel Weight changes.

The FMS has to be initiated prior to flight, following these steps:

- Enter zero fuel weight
- Insert center of gravity
- Enter block fuel
- Compare the trip fuel and flight time with the flight plan
- Compare the contingency fuel (route reserve)
- Ensure that reserve fuel is accurate
- Compare estimated take-off mass (ETOM) with actual take-off mass (ATOM)
- Check the landing mass

Inserting climb winds will improve accuracy, climb speeds, and fuel and time predictions. The winds will be used later by the FMS unless updated later at various waypoints. Climb winds can come from descend wind history or can be uploaded automatically.

Accurately programming the FMS is particularly important for long-range flights. It is essential to enter the winds and temperatures at each waypoint. A 1°C deviation from the Static Air Temperature (SAT) changes the true airspeed by 1 knot.

The FMS fuel predictions have to be checked while en route: arrival fuel at destination, fuel to alternate, the route reserve, the final fuel reserve (holding fuel), and the extra fuel and time.

Winds above and below cruise flight level have to be considered. The FMS systems use winds above and below the cruise flight level to determine the Optimum Altitude. The FMS will consider the cost Index when determining the Optimum (recommended Flight Level) altitude. Improperly programmed FMS will entice crews to add fuel to compensate for inaccuracies. This can be costly. On long-range flights, it can have an impact on payload.

Some flights are planned long before flight departure. Once airborne, the arrival fuel and Estimated Time of Arrival (ETA) have to be crosschecked with the flight plan, and any differences reconciled. It may take up to 4 h to upload the winds in the Flight Planning System. Updating the winds in long flights is important. When cruising at altitudes different from flight planned, the new winds and temperatures for the new flight levels have to be entered into the FMS. Finally, for improved descent profile management, the descent wind forecast has to be entered.

5.4 Tankering

Airlines normally use their aircraft in a continuous way, either in two-ways flights or jumping from one destination to another one. One of the most important decisions they have to take regarding fuel consumption is whether they upload, at a given airport, just the necessary fuel (including reserves) for the flight to the next destination, or if they include additional fuel for the following flights, reducing or eliminating thus the need to refuel in those points. There are several reasons that may justify this type of decisions:

- Large differences in the fuel prices between the departure and destination airports, making convenient to upload as much as possible in the cheapest point.
- Need to reduce the turnaround time at the destination airport, shortening or eliminating the time necessary to refuel.
- Supply problems at the destination airport, due to the lack of adequate aviation fuel, or its poor quality of contamination. There might be other type of restrictions: overprotection of local companies, limited fuel loading volume, or even the supply interruption at the airport due to a war situation, attacks, or accidents.

The inconvenience with this procedure is the increase in the aircraft fuel consumption and, at the same time, the higher maintenance costs. The extra uploaded fuel increases the aircraft fuel and therefore increases the fuel consumption. In the modern jets currently in service, any additional kilogram of loaded mass increases between 20 and 40 g (2%–4%) the burned fuel per flight hour. Then, increasing the uploaded fuel at an airport to reduce the load at the other one will be only justified from an economic point of view if the savings produced by any of the previously mentioned reasons exceeds the costs of the extra consumption (and the rest of operational costs).

This type of practices is generally known as tankering. There is no consensus about the concept being described by this term. In some cases tankering refers only to the additional uploaded fuel due to price differences, while in some other cases the expression covers all the cases when more fuel than the strictly needed amount is uploaded.

The amount of fuel that is authorized for tankering has to be added to the amount of fuel required by the flight plan, and it is limited by the aircraft certified weights: Maximum Take-off Mass (MTOM), Maximum Landing Mass (MLM), Maximum Fuel Mass (MF), and by the acceptable operating flights in the route, which is going to be flown.

In countries where the fuel supply is a government monopoly, the price of kerosene is normally uniform. There might be small variations coming from the application of local taxes. With the liberalization of the fuel distribution industry in many countries, for instance in the European Union, the fuel prices vary from one airport to another depending on a number of cost elements, which are relevant for the fuel selling price formation. The most important cost elements are the kerosene international price, transport to the airport, storage and supply to the aircraft. Additionally, the potential national and local taxes, the commercial costs, financial costs, and the profit margin for the supplying company have to be added. There might also be concessions from the supplying company to their best customers, normally always depending on the volume represented by their order. Another reason for fuel price differences is the trading on the futures market or hedging.

In those liberalized markets, airlines usually have tankering policies, especially for economic reasons, although the increase in the turnaround time at some airports caused by the excessive time taken for refueling also promotes this practice. Tankering is used generally in short- and medium-range flights, and very seldom in flights longer than 5 h, because in these cases the supplementary fuel consumption due to the additional fuel transported is too high. Airlines in the low-cost segment, trying to maximize aircraft utilization, may fill up the tanks in the first flight of the day and not do any refueling until it is needed to comply with the next trip flight plan.

The beneficial effects of this practice may be appreciable, although small compared to the total fuel expenses of the airline. In order to estimate those potential economic advantages, companies use simple mathematical models, like the one explained hereafter.

The transport of the additional fuel would cost:

$$\alpha_i T_i P_i + Cop_i$$

Being T_i is the tankering additional fuel uploaded at scale i; α_i is the additional fuel consumption per unit fuel caused by transporting the amount T_i in the flight between scale i and the following scale $i+1$; P_i is the fuel price at scale i; and Cop_i is the increase in the operational costs because of operating with a higher weight.

The term α_I depends exclusively on the aircraft-engine combination and the flight duration, and so does Cop_i. In small aircraft, one could think that an increase in the aircraft weight could affect the flight speed. However, in modern commercial jets, the cruise speed can be assumed constant. The aircraft take-off weight increases because the addition of T_i may require a higher engine thrust and an increase in the maintenance costs. In longer-range flights, the heavier weight may cause the initial cruise level to be lower than the optimal one. All these factors convey into a certain increase in the operational costs that may be ignored when evaluating the tankering economy.

The economic difference between the fuel brought from scale i, compared to the purchased at $i+1$, can be expressed as:

$$T_i(1-\alpha_i)(P_{i+1}-P_i)$$

The benefit obtained with the operation between scales i and $i+1$ would be the difference between the advantage and the cost of the fuel transport:

$$B_i = T_i(1-\alpha_i)(P_{i+1}-P_i)-(\alpha_i T_i P_i + Cop_i)$$

The amount T_i is limited by the aircraft design restrictions:

$\mathrm{OEW}_i + P/L_i + F_i \leq \mathrm{MTOM}_i$.
$F_i \leq \mathrm{MFM}$.
$\mathrm{OEW}_i + P/L_i + R_i + T_i(1-\alpha_i) \leq \mathrm{MLM}_{i+1}$.

where OEW_i is the operational empty weight of the aircraft at scale i; P/L_i is the payload of the flight between scales i and $i+1$; F_i is the flight on board before departing from scale i; $MTOM_i$ is the aircraft maximum take-off weight in the airport at scale i; MFM is the maximum fuel load admitted by the aircraft tanks; R_i is the fuel reserve necessary for the flight between i and $i+1$; and MLM_{i+1} is the aircraft maximum landing weight at the airport in scale $i+1$.

Most of the airlines require a minimum amount of profit in order to do tankering. As α_i increases with the flight duration, in long-range routes the profit margin decreases, and this tankering practice becomes less usual. This mathematical approach can be applied to a single-flight case, with a direct, easy solution and to a series of consecutive flights, with a more complicated linear programming result.

5.5 Redispatch and reclearance

The fuel reserves that aircraft have to upload are necessary in order to comply with the safety regulations of the airworthiness authorities, and therefore mandatory for every airline. It has been previously shown that fuel reserves are an important amount of fuel that is not consumed in most of the flights, because normally there are no contingencies. That extra weight however supposes an additional fuel consumption that, depending on the fuel price, may cost airlines a lot of money. In some long-range routes, it even reduces the allowable payload, leaving passengers and/or freight on the ground. This fact explains that airlines try to look for ways of reducing this extra cost without reducing safety.

Redispatch and reclearance procedures offer significant savings on long flights. Both are somehow similar, but redispatch technique is preferable because Air Traffic Control (ATC) clears the flight all the way to destination from departure and all the necessary fuel requirements are clearly established before flight. Reclearance, on the other hand, requires the flight to change destination while en route, which is cumbersome from ATC point of view.

The philosophy of the procedure is based in dispatching the aircraft with a flight plan to an intermediate airport and, close to that point, doing an additional flight plan from that intermediate airport to the destination. If there is enough fuel on board to satisfy that second flight plan, the aircraft continues to the destination, if not, goes to the intermediate point to refuel. The savings come from the fact that the addition of the two flight plans fuel is lower than the amount needed to fly to the final destination airport.

Although there are some differences in the various regulations, the reserve contingency fuel is normally a function of the flight distance or of the trip fuel. Reclearance allows for an uploaded reserve fuel reduction, decreasing therefore the total fuel weight and in this way, either increasing the payload weight (and consequently the revenue), or keeping the payload constant, decreasing the total aircraft weight, and so reducing the fuel consumption for that flight.

The practical application, following the European Air Safety Agency (EASA) regulations, is detailed here and can be seen, schematically, in Fig. 5.6.

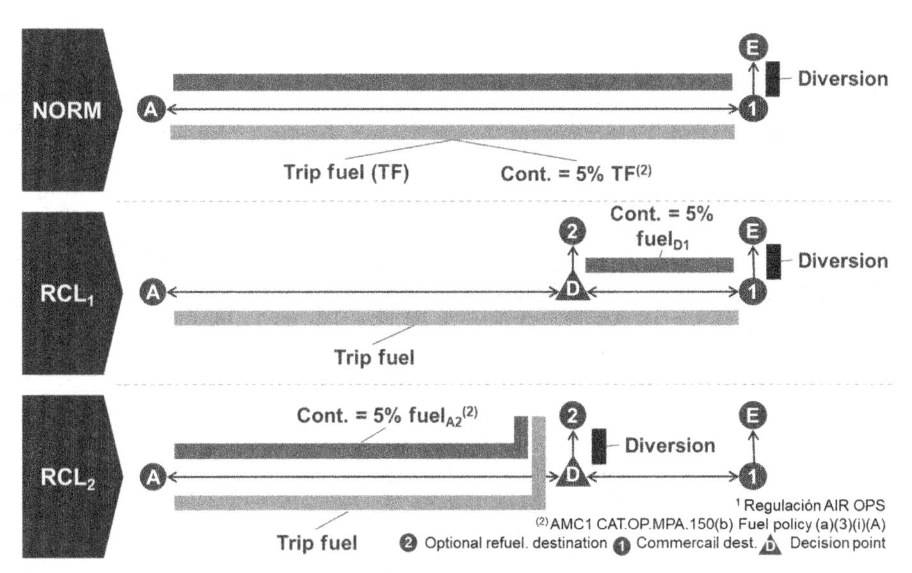

Fig. 5.6 Alternatives for reclearance (*EASA, AIR OPS regulation*).

The flight from A and the originally intended Destination 1 is divided into two flights: a first one from A and an airport Destination 2 a bit closer to A than Destination 1, and a second one from a so-called decision point in route, D (close to Destination 2) to Destination 1. When dispatching the aircraft in A, two feasible potential scenarios are calculated:

- RCL1: the 5% of the fuel necessary to flight from D to Destination 1 is calculated and added to the fuel necessary to fly from A to Destination 1.
- RCL2: the fuel necessary to fly from A to Destination 2 is calculated, and a 5% contingency added. When arriving at D, a new flight plan is determined to go from D to Destination1. If there is enough fuel on board, the flight continues. Otherwise, the aircraft lands on Destination 2 for refueling.

In a flight with reclearance, the uploaded fuel is the larger than the ones calculated in the two scenarios RCL1 and RCL2.

The consequences of this reclearance are:

- A reduction in the reserve fuel is achieved, translated into a smaller consumption and, sometimes, a larger payload.
- It requires an important precision in the calculation of both flight plans.
- It may increase the cost and damage the commercial image of the company if eventually it is necessary to fly to Destination 2 airport relatively often.

The optimum place for the intermediate destination is below the orthodromic of the route to minimize deviations and at a distance from A between 75% and 80% of the total distance. It can be seen that scheduled airlines use redispatch when there are payload gains or if the aircraft type has not enough range to reach final destination, with an acceptable payload, following the standard flight plan. Charter companies are

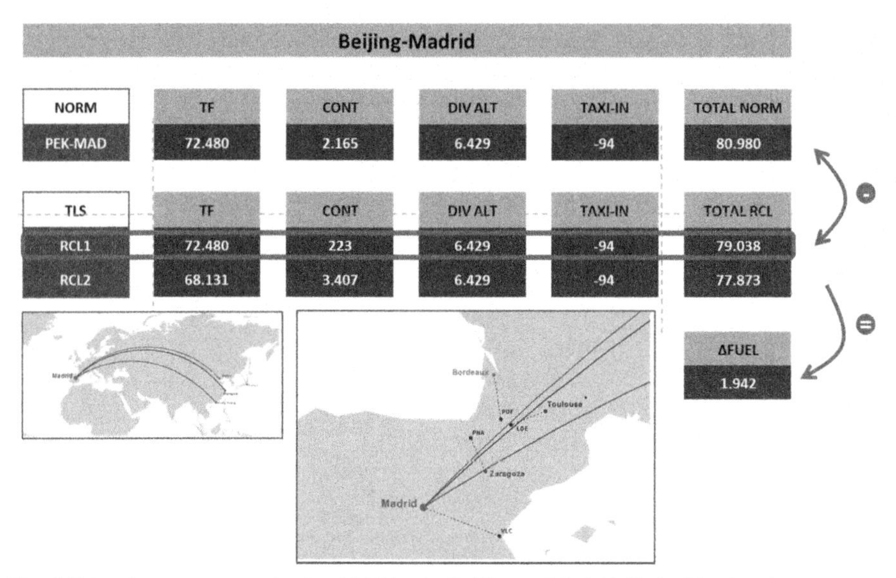

Fig. 5.7 Reclearance example. An A330 in the Beijing to Madrid flight (*Iberia, Operations Department*).

more frequent users of redispatch or reclearance because an unexpected intermediate stop is not so damaging for their commercial reputation.

An example of the fuel savings in a Beijing (PEK)—Madrid (MAD) route is shown in Fig. 5.7.

In this example, the intermediate airport 2 is Toulouse (TLS), well placed below the flight path, but a little too close to the destination Madrid. In the comparison between the original flight plan and the two reclearance alternatives, described in Fig. 5.6, the worse of them (RCL1) requires 1942 kg of fuel less than the basic flight plan. This can be translated to 19 additional passengers or the same weight in freight.

Operational procedures

6.1 Introduction

The basic rules for operating fuel savings according to the IATA Flight Path to Environmental Excellent document are:

- Program the most efficient aircraft for each route
- Choose the best taxi out way
- Fly the most efficient trajectory
- Fly at the most efficient speed
- Keep the most economic altitudes
- Maximize load factor
- Minimize aircraft operating empty weight
- Upload the minimum amount of fuel required by safety reasons
- Minimize the number of noncommercial flights
- Specific maintenance procedures for engines, airframes, and systems

Some of those rules are dependent on the airline flight planning and aircraft scheduling and others have been dealt with in the previous chapter when discussing the flight planning optimization. However, even having the best possible flight planning on hand, there are a number of important fuel efficiency decisions to be taken by the crew during the flight, with the help of the ground flight monitoring team. The best practices for those cases are the argument of this chapter.

6.2 Ground operations at departure

When the crew gets into the aircraft, the first decision to take related to fuel efficiency is the use of the APU or other source of energy until the engines are started. During the ground operations, APU can be used to supply electrical power and air conditioning to the aircraft. An APU high utilization implies high-fuel consumption and a costly maintenance. On average, APU fuel consumption (and the related emissions) is between 10 and 15 times higher than alternative ground power options. Consequently, managing the Auxiliary Power Unit (APU) is a great opportunity to save fuel and cost.

Although some old commercial aircraft types, like Boeing B-707 or Douglas DC-8 could be delivered without any APU, all modern models with a capacity greater than 50 seats have this equipment. Main reason is the difficulties by the airports to guarantee the availability of alternative ground power in all parking positions. Some airports can offer that but their number is relatively small, making convenient for airlines to have an additional energy source inside the aircraft in order to save more expensive engine time.

Energy Efficiency in Air Transportation. https://doi.org/10.1016/B978-0-12-812581-6.00006-5

A second but also important reason is the use of APU for ground operations, like turnaround servicing, a/c cleaning, cabin air-conditioning, etc., a series of activities that can require energy on board. The same consideration is applicable to maintenance tasks, like routine servicing, small repairs, checking electronic systems, and other inspections needing energy supply.

Establishing control and accountability of who and for how long to use the APU can result in difficulties. The APU use is not tracked at a user level, with multiple departments in the airline using it, typically with no procedures for transferring control and as a result no clear accountability for its utilization. Reducing APU utilization is a complex subject where many users within the company are involved. It requires extensive observation to determine how, where, and how long APU is actually used, and requiring crossdivisional cooperative effort.

A typical APU utilization allocation time is approximately 20% for Flight Operations, 10% for Maintenance Support, and 70% for Ground Operations. In all those functions there is potential for reducing APU utilization, but clearly the largest opportunities arise during Ground Operations and Servicing. There are many operational factors involved: the operating environment, the availability of fixed or mobile ground power units and air conditioning, the station turn time and ambient temperatures, the gate power compatibility with the aircraft systems, the parking locations, the ground support staffing level, the ground support service agreements, etc.

Aircraft consume a lot of fuel on the ground. Airplanes burn on average about 29% of the cruise fuel on ground (measured in terms of fuel flow, in kg/h). A number of good practices recommended by IATA can help saving fuel from the very first phases of the flight.

In case of delay at departure, it is preferable holding at the gate, delaying the pushback and absorbing some of the delay at the gate with the engine off. The tow bar and communication cord should be disconnected as soon as possible after brakes are set. To minimize the power requirements during the initial roll out and minimize ground hazard, the aircraft should be positioned in the initial taxi out direction. An engine-out taxi procedure should be considered when the ramp and taxiway conditions permit, or in case the aircraft weight is below maximum landing weight or if the anticipated taxi time and specific aircraft system permit. When unanticipated delays are encountered during taxi out, engine out taxi or shutting down engines should be considered during extensive delays.

If a flight's weight is light, and the flight crew chooses to taxi with all engines running, it may be necessary to make constant use of the brakes. This can cause excessive wear and heating of the brakes. Engine out taxi requires slightly more anticipation compared to taxiing with all engines operating. Before using engine out procedures, airlines must develop proper standard operating procedures (SOPs) and properly train crews.

Taxi speed is important. A lot of time can be gained or lost while taxiing. Flight crews must remember that the fuel burn with engines that are idling on the ground equates approximately to 25%–30% of cruise power. In ideal conditions, the recommended taxi speeds should be around 10 knots for maneuvering. On straight taxiways, speeds up to 30 knots are acceptable.

6.3 Take-off and climb

Take-off, initial climb, and climb to initial cruise level are very critical phases of the flight where an important amount of fuel can be saved depending on how operational procedures are followed.

Reduced thrust will not improve fuel consumption during take-off, but it will preserve the engine deterioration, which ultimately reduces fuel consumption over time. In a modern turbofan, the first 1% reduction from full take-off thrust will result in a 10% saving in engine life. Consistent use of reduced thrust will more than double engine life and prevent a rapid performance deterioration. In addition, the nitrogen oxides emissions (NOx) of the engines are growing exponentially with the engine combustion chamber temperature. Engine derate at take-off makes the engines work cooler and drastically reduces the emitted amount of this contaminant.

If airport configuration and weather allow, it is recommended not to overuse the high lift devices and to select the flaps minimal deflection complying with the requirements of the corresponding performances. In this way, the aerodynamic drag is reduced, the climb performances are optimized and the aircraft spends a shorter time at lower altitudes, consuming therefore less fuel. The use of lower flap setting will improve the second segment climb and lessen the flap and slat clean up time thereby reducing the fuel burn.

An example of the fuel savings due to the use of lower flap setting can be seen in Table 6.1 with four Boeing aircraft, two medium range (B-717 and B-737/800) and two long range (B-777/200ER and B-747/400). Savings go from 10 to 64 kg per take-off. It is interesting to note that short-range aircraft make more flights a day and the savings of 4 B-717 take-offs flying 2 h average stage length trips are equivalent to a single B-747/400 flying an 8-h trip.

It is also recommended when taking off to retract flaps, slats, and landing gear as soon as possible. Although the flaps and slats increase lift, they also increase drag

Table 6.1 **Take-off fuel depending on flap selection** (*Boeing AEROMAGAZINE*)

Aircraft type	Take-off flap setting	Take-off gross mass (kg)	Fuel used (kg)
	5		423
B-717-200	13	51,256	431
	18		438
	5		578
B-737/800 (Winglets)	10	72,575	586
	15		588
	5		1635
B-777/200ER	10	249,476	1668
	20		1692
B-747/400	10	328,855	2555
	20		2618

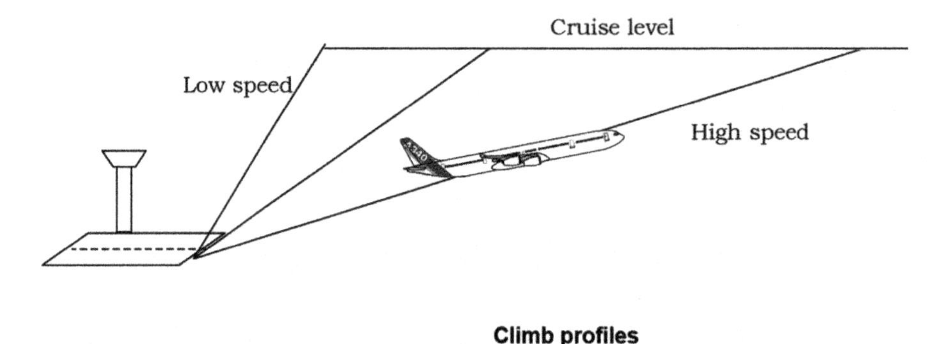

Fig. 6.1 Different climb profiles (*Airbus: Getting to grips with fuel economy*).

and therefore increase fuel consumption. When departing in a direction opposite to the desired en route course, the take-off flaps should be maintained, trading speed for altitude until the aircraft reaches the initial altitude where a turn to the on-course can be initiated. Maintaining a lower speed will allow for a faster turn rate to the on-course for a specific bank angle (when possible bank angles of up to 30° should be used).

The slope of the climb to the cruise altitude has a big impact on the fuel consumption, and obviously also on the flight duration. The crew may decide going faster and with a lower climb slope or using a lower speed and a higher climb slope, as indicated in Fig. 6.1. A number of airport conditions, like the existence of obstacles or noise abatement regulations, may limit the freedom of choice by the crew.

In the absence of other limitations, the Cost Index can help optimizing the climb speeds. Table 6.1 shows the effects of different CI in the climb up to cruise level 330 (33,000 ft.) of two different aircraft models: a medium-range A320 and a long-range A330. The applied CI goes from 0 (no time value, minimum fuel burnt selected) to 200 (moderate time value, the maximum is 999). The A330 is faster thanks to its better thrust power ratio and the time difference between the low and high CI options is less than 3 min and 30 nm, but fuel consumption increases 500 kg in the fastest mode. The A320 time difference is 1 min more and 37 nm, but consumption only increases 300 kg (see Table 6.2).

The continuous climb departure (CCD) procedure, avoiding steps, allows for a fuel consumption reduction and shortens flight times, while it reduces also the number of ATC authorizations. This procedure is still in its infancy and presently limited to a number of airports with specific local procedures. In any case, it needs to be tailored to the specific operative runway. Fig. 6.2 shows an application to Dallas Fort Worth airport, in which a mandatory temporary level off is included (Fig. 6.2).

On short flights, the most efficient regime from the energy point of view is a continuous minimum fuel climb until the aircraft trajectory intercepts the optimum descent profile. The practical difficulties associated to this procedure affect to ATM, as the aircraft crosses many flight levels without stabilizing in any of them, and from the onboard service, not having any cruise period to perform its duty. Most flight planning systems normally assume a minimum cruise time of 5–10 min in determining the optimum altitude. The total air distance should be considered when selecting the optimum altitude on short flights, including the departure and arrival runways and procedures.

Table 6.2 Minimum cost climb profiles climb profiles according to different Cost Index values (*Airbus: Getting to grips with fuel economy*)

Aircraft type	Cost index	Fuel (kg)	Time (min)	Distance (nm)
	0	1757	22.4	150
A320-200	20	1838	23.1	159
CFM56 engines	40	1897	23.7	165
75-ton MTOM	60	1980	24.7	175
	80	2044	25.6	183
	100	2080	26.1	187
	0	3568	19.1	122
A330–200	50	3773	20.0	135
PW4168 engines				
200-ton MTOM				
	80	3886	20.5	141
	100	3927	20.7	143
	150	4005	21.3	148
	200	4068	21.7	152

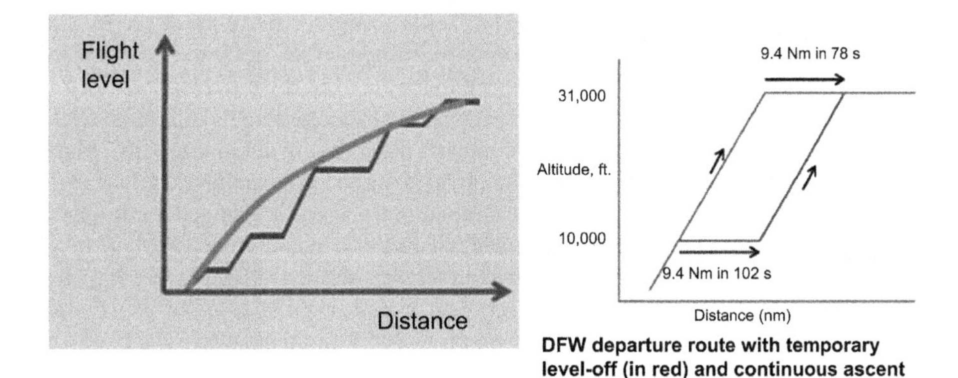

Fig. 6.2 Continuous climb departure procedure (*DFW airport web page*).

6.4 Cruise

As discussed previously in Chapter 5, the vertical profile management has a big impact on the flight's fuel efficiency. An accurate flight planning system will produce the best vertical profile based on the aircraft weight, the wind field at each waypoint, temperatures, and the flight specific Cost Index. This is assuming that the airline is using the correct Cost Index values and that the flight planning system incorporates those Cost Index values in its vertical optimization process. Flight planning systems normally check all available altitudes up and down to achieve the minimum cost per ground mile. This may include descents to lower altitudes to take advantage of better wind and TAS (true airspeed) combinations.

The process of selecting the optimum flight level can be shortened using a database of historical references for that flight. With this input, the system will start checking the levels with the highest probability of being adequate and reduce the total number of calculations.

Efficient flight planning systems should consider all possible routes or portions thereof to determine the most efficient routing, including a number of special circumstances like:

- Planned departure runway and applicable procedures that may change the trajectory of the beginning of the flight, requiring a turn to aligned with the route.
- Winds and temperatures at all altitudes. Airlines use to have some statistical rules for the standard flight plan, using 50% probability winds (the wind intensity having a 50% probability of not being worse) in short-medium range, and 85% probability in long range. Actual winds will replace those standard data as required.
- Cost of air navigation charges in the different possible flight tracks.
- Airways temporary restrictions, NOTAMS, status of restricted areas.
- Arrival runways and expected arrival procedures.

The problem with many flight-planning systems is that route analysis is based on a fixed Mach number instead of a Cost Index optimized vertical profile. The optimum profile will be flying a continuous ascend trajectory as the aircraft is burning fuel and losing weight, modifying speed accordingly. This is not appreciated either by ATM who try to stabilize the aircraft at one of the approved flight levels, or by the Commercial Department who prefer flying constant speed to have a fixed flight duration.

The compromise solution used to be a stepped cruise, reaching the optimum flight altitude after the climb and stabilize the aircraft there, flying at close to LRC Mach number. When the aircraft gets lighter enough to go up to the next allowed cruise level (see Chapter 8 for details on the airspace structure), the speed is changed according to the new altitude conditions. As already seen, this is the standard procedure, but actual wind and temperature conditions may give preference to certain flight levels, not necessarily the highest.

The objective of lateral track optimization is to find a minimum cost route based on the planned or tactical Cost Index, looking at all possibilities vertically and laterally. Very high Cost Index values will tend to drive the aircraft down toward the crossover altitude due to the higher TAS values at those altitudes.

A factor that is not generally included in the cost index calculations is the price of the air navigation charges in the different flight tracks. The price differential may be very high. Just looking at the European airspace, the most expensive charges (Switzerland) are almost 10 times the cheapest (Portuguese Azores FIR) and this can become an important element in the achieving a minimum cost flight plan (see Chapter 3).

Crews should attempt to fly the planned track while taking short direct routings to minimize large turns at waypoints. When accepting a long direct routing, there is always a danger of crossing restricted, military areas, high navigation cost areas (particularly in the EU), or bad weather. On long flights the route should be re-evaluated, because after several hours of flying the wind forecast might have changed. Route changes however are rare as they are complex (i.e., Pacific routes). If possible, predetermined routes and fixed altitude should be avoided. The route should be optimized

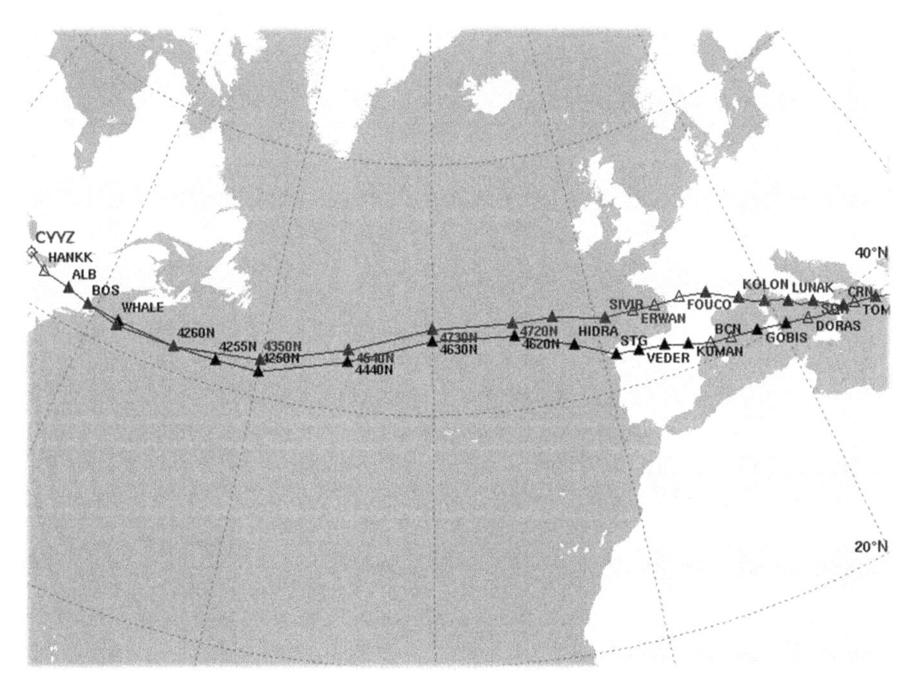

Fig. 6.3 Comparison between minimum fuel (upper track) and minimum cost (lower track) (*IATA Training: Fuel efficiency and conservation*).

according to the flight conditions for the day of operation, except for heavy traffic, specific local procedures or restricted by a preferred ATC route system (RADs), delays from altitude restrictions.

Fig. 6.3 compares two routes in the North Atlantic. The one more to the North is the result of computing the minimum fuel consumption and the Southern one, the minimum cost. Among the different factors to be computed are the aforementioned differences in air navigation charges.

In Chapter 5, the two key parameters to select the flight regime, i.e., the flight speed and altitude, were discussed. Speed is probably where the airline has more decision capacity, since the flight levels are many times assigned by the air traffic control (ATC) and airlines have far less margin of maneuver. Aircraft have several characteristic flight regimes. The most common ones are the maximum range cruise (MRC) and the long-range cruise (LRC). The MRC consists of flying at the Mach number that maximizes the specific range, i.e., the distance traveled per unit fuel consumption. The LRC consists of flying at a slightly faster speed, so that the specific range is 1% lower than for the MRC. Non-FMS equipped aircraft use LRC speeds. The specific range curves as a function of the Mach number for each aircraft type depend on the aircraft weight and the flight altitude. A graphic representation of these two regimes can be seen in Fig. 6.4.

Increments in the Mach number above the MRC result in increases in the flight consumption without significant reductions in the flight time. For instance, in the case

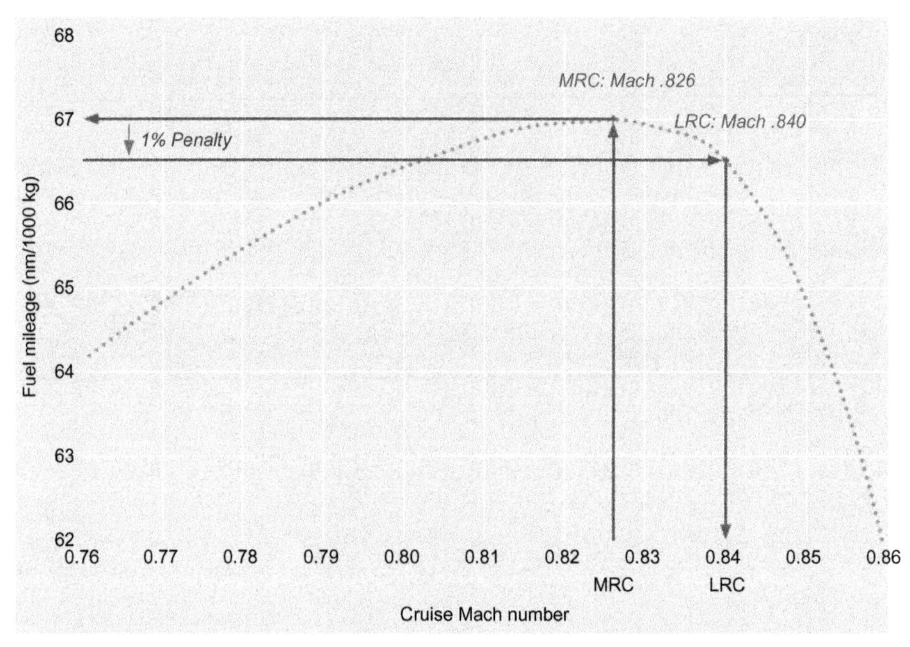

Fig. 6.4 Difference between Long-Range Cruise and Maximum-Range Cruise (*Boeing: AEROMAGAZINE Q4-07*).

of Boeing aircraft, depending on the model, a Mach number increase between 0.03 and 0.04 reduces the flight time in a 5000-mile cruise between 20 and 30 min, with fuel consumption increases between 6% and %, always with respect to the MRC.

Several time-related costs occur during the day-to-day operation, which justifies the use of Cost Index to develop new versions of the basic flight plan:

- Strong headwinds at the selected flight levels may result in costly delays
- A last-minute delay to wait for some connecting passengers
- Impact of the forecasted delays on subsequent flights
- Curfews or slot times at destination airports
- Gate occupancy conflicts
- Crew routing problems as they are losing connections as well, changes in limits of activity by long delays, asking for a reinforced crew or do not allow to continue flying

This is normally call "tactical" CI. The use of tactical CI values will minimize the time-related costs even though additional fuel could be consumed in the process.

Flight crews are normally in a difficult position to decide on the most appropriate Cost Index for the flight. Flight Dispatch, based on input from Operations Control, should proactively monitor the flight process. To improve flight management and reduce operational costs, airlines must ensure that all processes are well defined, managed, and fully integrated.

Dynamic CI will minimize overall costs and improve on-time performance (OTP). CI utilization will result in substantial fuel and/or time savings, while balancing the time and fuel costs based on the specific airline cost structure (see Chapter 5). Unless requested by Flight Dispatch/Ops Control, the FMS-managed speeds should not be overridden. If the ETA varies from the normally scheduled arrival time, coordination is required with Operations Control and dispatch to adjust the ETA.

An early arrival situation should also be coordinated with Dispatch/System Control. A lower-than-standard Cost Index can reduce fuel costs, prevent possible gate holds, reduce the chance of ramp congestion, and improve ground staff costs. While the cost of fuel should be minimized, all time costs must be considered when selecting a specific mission Cost Index. A pre- or postdeparture reoptimization of the flight speed profiles should be considered to reduce time-related costs (early or late).

Flight altitude is also a very important parameter determining the fuel consumption. Cruise curves for each aircraft type indicate also that there is an optimal altitude for which the specific range represented as a function of the flight altitude shows a maximum. This optimal altitude depends on the Mach number and the aircraft weight. Since the aircraft weight reduces as the flight progresses as a consequence of the fuel consumption, the optimal altitude changes also in a continued way (increases as the weight reduces). The optimal cruise would be therefore a continued ascent, just limited by the operational ceiling of the aircraft.

ATC does not currently permit this type of trajectories, so in practice it is impossible to fly constantly at the optimal altitude. For long enough flights, the most approximated trajectory is the stepped cruise, where the flight level increases step by step, normally of 2000 or 4000 ft. According to Boeing information (Boeing AERO Magazine), for a 4-h cruise, flying at an altitude, which is the average of the optimal altitudes of the whole 4-h cruise, implies a 0.6% extra fuel consumption. This extra fuel consumption becomes 4.8% if the aircraft flies 4000 ft. below that altitude.

The effect of wind, both head wind and tail wind, is very important with regard to fuel consumption and may also affect the selection of the flight level. In the crew operations manuals, there are charts allowing them to determine the best flight altitude depending on the forecasted wind speed.

The previous chapter described how computerized flight plans have started to allow the introduction of the last-minute data relative to the aircraft occupation and loading of that particular flight, as well as meteorological data. It is the so-called flight management system (FMS) that is able to optimize in real time the aircraft performance, improving the flight economics, not just in terms of fuel consumption, but also improving the operational costs. For instance, it allows for a better control of the aircraft centering. Keeping the aircraft centered in the right position permits flying with a smaller aerodynamic drag, and therefore less fuel consumption, compared to flying with, for instance, a moved-forward position of the aircraft center of gravity (CG). The more aft the CG, the less induced drag is produced. This improves the specific range of the aircraft due to fuel burn improvement. Depending on the aircraft type, drag created by loading an aircraft to the forward limit of its forward CG can increase drag by as much as 3% compared to loading the aircraft at its most rearward center of gravity limit.

One must not neglect fuel consumption as a consequence of the aircraft heating and air-conditioning systems. It is recommended to the crews to use these systems within reasonable temperature limits. This system is typically responsible for between 0.5% and 1% of the fuel consumption of a flight.

6.5 Descent and approach management

A properly planned and executed descent profile offers interesting opportunities for fuel savings. The ideal profile would be an uninterrupted descent from cruise altitude without the use of thrust or speed brakes until reaching the final approach stabilization altitude. The flight crew should adhere closely to the computed descent speeds and monitor the descent profile to determine early if adjustments are required. If it matches this profile, the correction should be increasing the speed rather than using speed brakes. If not, the correction should consist of reducing speed slightly to regain profile or make power adjustments for profile correction as high as possible.

The FMS will calculate the top of descent (TOD) as a function of the Cost Index. The higher the descent speed, the steeper the profile, the shorter the descent time, and the later the TOD. Contrary to climbing, the landing weight and the TOD fight level appear to have a negligible effect on the descent speed computation.

The FMS computes accurate and efficient descent profiles. Except for tactical reasons, flight crews should not intervene by descending early or late, or modifying speeds or rate of descent. Unless the profile is modified by ATC, it should be flown as computed. It is also recommended not to change to the landing configuration too early, avoiding taking flaps early and keeping aircraft clean.

Descent planning and energy management are critical. The FMS should be programmed for the anticipated descent and approach pattern. Otherwise, the TOD point will be erroneous and the aircraft's energy state will be incorrect. Poor FMS programming may lead to a destabilized approach. The CI optimizes the descent speeds.

During descent, there is a tradeoff between speed and attitude. The FMS is continuously working toward the next altitude and/or speed restriction. On descent, if the flight encounters a temporary altitude restriction that takes the aircraft high on the profile, the speed, unless a hard speed is assigned to ATC, should be reduced as much as possible down to minimum drag clean speed if necessary. Then the speed can be subsequently increased back to normal descent or higher speed if required to regain the descent profile.

Some trajectories, designed to minimize fuel consumption at landing such as continuous descent approach (CDA), are more and more common. According to EUROCONTROL definition, a CDA approach procedure is an operational flight technique where the aircraft descends in a continued way, from an optimal position, with minimum thrust, avoiding the typical steps in the approach procedures, complying with the minima established in the published procedures and the ATC instructions in order to guarantee a safe operation (Fig. 6.5). The objective is to fly optimal descend profiles, reducing the fuel consumption and the emissions.

CDA was initially thought for reducing noise in approach, with a significant impact between 15 and 35 km from airport, although lower at shorter distances. But CDA also reduces emissions and fuel consumption. The problem is that it requires ATC changes and may limit the airport capacity, reducing the maximum number of operations per hour.

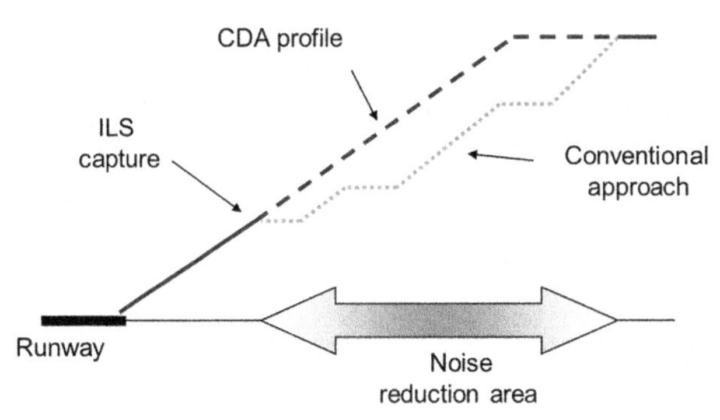

Fig. 6.5 CDA profile (*IATA: Aviation and the Environment*).

In high-traffic activity situation, the capacity of the airport is limited by the separation between approaching aircraft. When CDA technique is used, approaching aircraft are flying low thrust, minimum lift surfaces deployment. In this configuration, a sudden wind change can increase or decrease the speed of the aircraft, taking it closer to the previous or the next aircraft, with relatively short time to recover its position. Due to this reason, airport ATC may increase aircraft separation when CDA mode is being used.

An increasing number of airports are applying CDA although some of them are introducing some limitations, like having it only in low-traffic periods or doing it only in one of the runways while the others use the standard approach. There is some research about how to improve CDA procedures, looking at the possibility of exchange speed for minor altitude changes without getting closer to other aircraft (Fig. 6.6). The implementation of this procedure would need to modify FMS software.

Descent winds management is also important for fuel savings. For accurate FMS computation of the descent profile, the descent winds should be entered. Without

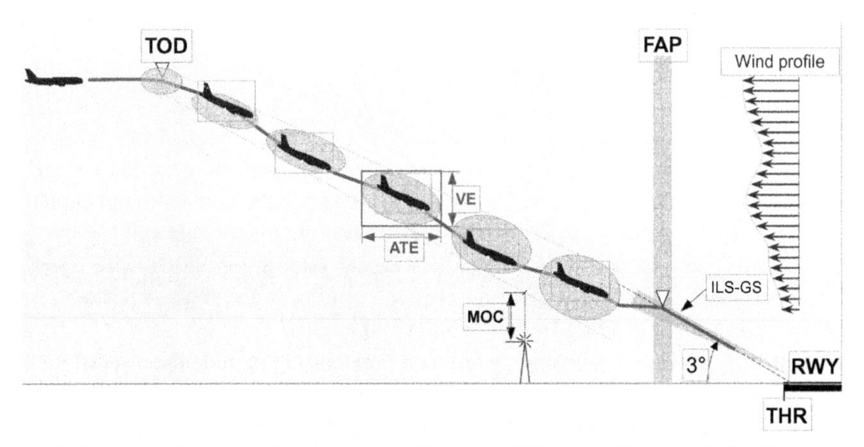

Fig. 6.6 Compensation of wind changes by altitude modifications (*Román, J. E.: Commercial aviation and the Environment*).

winds, a wind profile will be built assuming a constant decreasing wind speed from the cruise level down to the airport altitude. The computed descent winds are found on the FMS flight plan page and can be compared to the forecast winds. The forecasted winds should be updated in the FMS if different.

Generally, the following rules can be used for a quick calculation for distance, speed, and altitude tradeoff:

- In a clean configuration, level flight, and idle power, an aircraft will decelerate 10 knots per nautical mile, i.e., 60 knots speed loss will require about 6 nm.
- In a clean configuration and at idle power, in still air, most aircraft will normally descend 1000 ft. per 3 nm.

Early or late descent profile causes a significant increase in fuel consumption. If the descent is started early, the opportunity of regaining the optimum profile is possible and it should be done as high as possible. In the case of a late descent, the fuel is already consumed in cruise and is lost. The extra energy must now be dissipated with increased drag. Speed brakes should not be a substitute for adequate descent profile planning and proper descent profile management.

After the descent the approach phase begins, also with opportunities for fuel savings. Among them is the low noise/low drag approach, which was used initially to minimize noise in Europe. The advantages of the low noise/low drag (decelerated) approach are:

- Lower fuel consumption and emissions
- Lower noise levels
- Time savings because of higher approach speeds
- Flexibility and ability to vary speed to suit ATC

Another advantage of using the low noise approach is that it sets some clearly defined altitudes and speed targets to achieve during the approach. Most approaches at various airports will be completed in an identical manner improving thus standardization.

6.6 Landing and taxi in

Even at landing, fuel-saving opportunities may be relevant. Reduced flap landing for instance offers opportunities for fuel savings. Most airplanes are certified to land using less-than-full landing flaps, and some aircraft types even have autoland capability using reduced flap. At the last flap setting, more drag than lift is normally generated. Reduced flap landings will therefore reduce fuel consumption along with chemical and noise emissions. When landing an airplane with reduced flaps, fuel burn is reduced by approximately 15 kg in fuel on an A320 landing.

Considerations when performing a reduced flap landing include the aircraft landing weight, the runway length, the runway exist point and occupancy time, the runway surface conditions, possible tail wind component on final approach, and brake cooling during short turnaround times. According to IATA, the average increase in speed for reduced flap in landing is approximately 5 knots and the extra landing distance around 500 ft. Several airlines have made reduced flap landing procedures a standard.

Another way of saving fuel during landing is the engine idle reverse. Additional advantages of using idle reverse on landing include reduction in fuel consumption, reduction in noise and emissions, better passenger comfort, elimination of a high-power cycle on the engines, reduction in potential engine stall and reingestion, reduction of foreign object damage (FOD), increased engine reliability, lower cooling time requirements before shutting engines down for engine-out taxi, and slower engine performance deterioration.

Some airports have prohibited the use of thrust reverse at night due to noise abatement reasons, with the exception of safety considerations that need to be adequately justified after the landing. In order to quantify potential savings, 1t is important to notice that full reverser flue flow is about twice the cruise fuel flow.

From the economic point of view, there may be a concern on brakes wear due to the utilization of idle reverse, because brake maintenance is an expensive item. Steel brakes maintenance is a function of the number of landings and the time and pressure applied on them by the crew. Then, a tradeoff between brake refurbishment cost and reverse use can be done.

Things are different with carbon brakes because there is evidence that carbon brake wear is more a function of the number of applications rather than the amount of braking used, then the use of more or less reverse has no economic repercussion in this item. It goes without saying that the use of idle reverse on landing is always subject to safety limitations. Some safety factors should be considered, like runway length and surface condition, aircraft landing weight, tailwind on final approach, and turnaround time.

Once the aircraft landed, performing the taxi in operation with just one engine running (engine-out taxi) is nowadays a standard procedure in many airlines under normal conditions. The main advantages of this procedure are reductions in fuel consumption and consequently on engine emissions. Also, brake wear is reduced.

Of course, in order to facilitate compliance and keep the operational safety margins, appropriate SOPs are required. The procedure requires certain precautions: for instance the aircraft must be kept moving, therefore anticipation is important. Flight crews will require training and familiarization with engine-out taxi procedures, but the experience shows that crews familiar with engine-out taxi in procedures normally follow the engine-out taxi procedure after almost every landing. The procedure is not always possible and the taxiway surface condition has to be checked in advance, as well as the engine configuration. Other important considerations that may be relevant when deciding on using engine-out taxi in are the taxi in time, potential ramp congestion, and local airport regulations.

6.7 Summary

In addition to all the previously commented practices and procedures, IATA also recommends to airlines to take all necessary measure in order to reduce aircraft commercial equipment weight: catering carts, galley equipment, duty-free carts, cargo containers, pallets, stoppers, and nets. Some of these considerations are fully explored in Chapter 7.

Summarizing the operational measures mentioned in this Chapter, the example of Iberia can be brought in, with the company announcing a plan aiming at reducing fuel consumption based on the following operational measures:

- Aircraft weight reduction (reassignment of nearest alternate airports, aircraft interior renewal with lighter seats and trolleys, potable water load control)
- Cruise speed adjustments, flights at optimum altitude levels, engine-out taxi
- Optimization of landing maneuvers (engine idle reverse, lower flap setting, CDA)
- Maintenance: increasing the frequency of engine and fuselage wash, new and lighter aircraft painting
- Improved energy efficiency through flights planning incorporating Cost Index, fleet assignment for different routes, and load factor optimization

In particular, and based on the application of the actions described in this chapter, an airline can achieve a 2–3% fuel savings applying operational measures all along the flight:

- Flight planning: match planned and real operation; fuel optimized payload, route, Flight Level, and speed
- Execution excellence: best fuel efficiencies practices applied in flight execution; continuous feedback between Flight Ops Management and Crews, focusing on efficiency issues
- Balance weight on board versus profit: optimization of in flight retail and pantry as a function of profitability; overall cabin weight reduction: lighter trolleys, water, and magazines

Maintenance

7

7.1 Introduction

Since some decades, commercial aviation has become the safest and most reliable transportation mode, independent of the statistical parameter used to measure these features. The statistical trend shows a clear improvement trend and 2017 has been the safest year ever in commercial aviation up to the point that no passenger was killed in any commercial jet accident. Even admitting that it was an extraordinary year and there will be some accidents in the next future, it is a great success that gives a strong recognition to the commercial aviation safety system.

A basic pillar for this achievement is the instauration of a continuously updated robust maintenance system, the basic principles of which are, with slight differences, honored by all States across the world.

Before entering into service, every new commercial aircraft type must have obtained a Manufacturer Maintenance Program (MMP), approved by the Civil Aviation Authority (CAA) of the design State that remains as the minimum requisite for all the prospective operators. The airlines may add more actions to the program and reorganize their process sequence, according to their specific aircraft scheduling and maintenance organization, always with the approval of the aircraft registry CAA. The airline Maintenance program of a determined type of aircraft is exclusive of this airline and will not be applicable if the aircraft is sold to other operator.

A commonly used definition of commercial aircraft maintenance describes it as the execution of all the needed tasks for keeping the aircraft in optimum conditions from the technical, operational, and economic point of view. It has a great relevancy due to its participation in the operational success within an environment of heavy requirements about safety and operational reliability, in a highly regulated field, and being applied on very costly elements like commercial aircraft.

The main purposes of any maintenance program, ranked by level of importance, are:

- Ensure the correct safety levels of the aircraft operation, minimizing the probability of any accident or incident.
- Demonstrate the accomplishment of the CAA requirements, being able to pass successfully the corresponding inspections and retaining the individual aircraft Airworthiness Certificate.
- Maximize the time in which the aircraft are available for commercial service and, in this way, collaborate to increase the potential utilization of the fleet.
- Optimize the airline operating economy, taking into account all the relevant factors and not only those affecting the maintenance cost.

The two first goals are paramount and each individual aircraft has to be the owner of an Airworthiness Certificate where the registration CAA accepts that this aircraft operation is safe and in agreement of all the national and international requested conditions.

Energy Efficiency in Air Transportation. https://doi.org/10.1016/B978-0-12-812581-6.00007-7

To that purpose, the airline has to set up a Technical Organization with the responsibility of guaranteeing the practices and procedures for the compliance of all demanded maintenance standards. Both Technical Organization and its procedures are approved and audited by the corresponding CAA, being a key factor for the process of granting the AOC (Air Operator Certificate) to the airline. At the same time, the Technical organization needs to be also certified by the Type certificate CAA, as a requisite to sell that aircraft-type maintenance.

The optimization of the airline operating economy depends on many different factors. From the perspective of the capital cost, maximizing the utilization of the aircraft (generally the most expensive economic activity in an airline) is a basic element for improving operating economy. The majority of the airlines intend to operate their fleet at the maximum feasible number of flight hours, scheduling maintenance tasks in the low-demand period or leasing out its capacity surplus. The three key factors for that goal are a good aircraft routing, enhancing the time continuity of the flights, an adequate ground-handling operative capable to perform fast turnarounds according to the particular flight needs, and minimum maintenance periods, in such a way that maximizes the time in which the aircraft is ready to enter into service.

At the same time, an optimum technical aircraft condition will reduce fuel consumption with its logical repercussion in the airline's budget. The degradation of the aircraft performance during its service life may be compensated or mitigated by additional maintenance actions other than those requested by safety considerations. In many cases, these actions are going to take more grounded aircraft time and will reduce its availability for commercial service.

The balance between these opposed procedures is one of the first targets of the airline maintenance system and needs of special attention by a multidepartmental management group, to foster the consideration of global airline economy over the particular budget of individual airline units. For example, more intensive repair actions may need to extend the aircraft stay in the maintenance zone in order to perform different types of checks and put pressure on the airline schedule or even reduce the fleet utilization. On the other hand, minimum required maintenance work might reduce reliability, through more flight cancellations or delays, and increase fuel consumption due to higher airframe, engines, and systems degradation.

7.2 Maintenance and performance monitoring

In order to assess the aircraft energetic efficiency, engineering and maintenance departments keep a continuous monitoring of the specific consumption and other related parameters along the service life. The analysis of the specific consumption degradation may indicate possible technical deficiencies that can be repaired through maintenance actions or even nonoptimized operational procedures to be corrected by the corresponding airline department.

The control of the aircraft performance begins in the definition of the aircraft specification that has to incorporate the equipment needed for the performance monitoring, according to the operator requisites. Many operators inspect the aircraft final assembly to control adjustments, sealing and painting status, because the potential need of repairs and modifications before delivery may increase the aircraft weight.

Predelivery testing provides the data for establishing the aircraft performance baseline. The tests include fully equipped MEW check and a flight test to prove that the specific range of the aircraft is within the agreed limits and complies with the delivery guarantees in the contract. Other tests, not related with performance, examine different systems functionality or cabin related features, as In-flight Entertainment System (IFE) or cabin interior noise.

After the aircraft enters into service, performance monitoring can be done at fleet level (analyzing the efficiency trend of all the aircraft of the same type together), at individual aircraft level (establishing the trend of each individual tail number), and at serial number engine level (monitoring the performance degradation of each engine). Individual tail number and engine monitoring are more commonly applied to long-range fleets, when additional fuel consumption may be translated not only to the need of buying more fuel but also in potential payload reductions with heavier economic penalties. Short/Medium-range fleets do not suffer payload limitations due to fuel quantity in the majority of the cases and their performance can be controlled at fleet level or in groups of aircraft of similar age and specification.

Fig. 7.1 shows an example of the degradation evolution using a completely fleet trend analysis, while Fig. 7.2 indicates the same effect in an individual engine.

Airplane Performance-Monitoring (APM) procedures are based on the comparison of actual specific fuel consumption, generally in the cruise phase of the flight, with a baseline performance. This baseline may be the condition of the aircraft or engine at delivery or at the moment when the performance trend analysis begin. Typical parameters to be analyzed are:

- Volume of fuel burnt
- Energetic efficiency in terms of some selected indicators (fuel per RTK, fuel per flight hour)
- Degradation of the specific fuel consumption
- Adjustment of the actual performance with the predicted trend

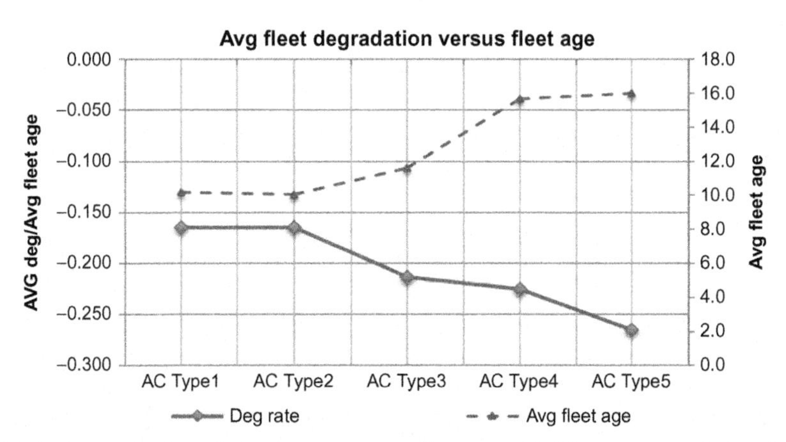

Fig. 7.1 Performance degradation versus average fleet age of 5 different types of aircraft (*IATA Training: Fuel efficiency and conservation*).

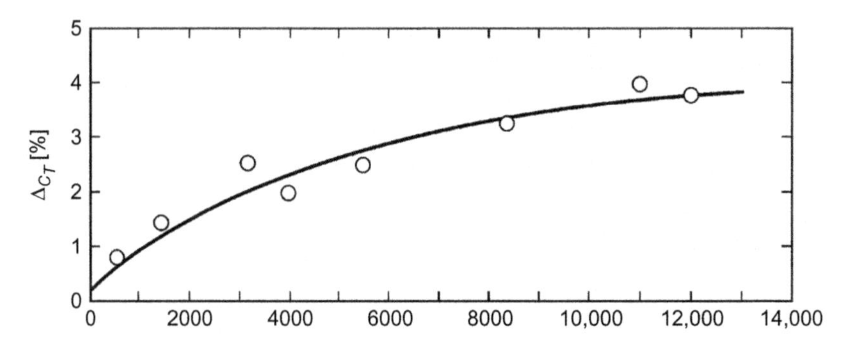

Fig. 7.2 Specific fuel consumption increase for a typical commercial jet engine as a function of the number of service hours (*Benito, A.: Technology and maintenance for fuel saving*).

The aircraft fuel consumption performance at delivery is checked during the predelivery flights. It is slightly different in each individual ship and used to be somewhat better than those reflected in the Airline Performance Manual (APMA), due to the conservative margins introduced to comply with the delivery condition guarantees the airplane manufacturer provides to its customers. If the airline is requiring additional in-service performance guarantees to the manufacturer, an additional application of the APM is to check whether the performance evolution of the aircraft complies with those guarantees. Guarantee conditions are inside the Aircraft Purchase Contract and have legal value. Therefore, all the APM technical procedures must be accurate and transparent, with the purpose of being used in case of a potential legal dispute. The high technical complexity of those calculations made the in-service performance guarantees very difficult to enforce.

The data needed for the APM process are recorded by on-board instruments, sent to the Digital Flight Data Recorder (DFDR), and communicated to the Technical Department. Depending on the aircraft's configuration, the transfer of data is done through printouts from the cockpit printer, extracted via CD or transmitted via Aircraft Communication Addressing and Reporting System (ACARS). The result of the analysis is used by the Operations Department to correct the fuel consumption relevant elements of the Aircraft Performance Manual and by the Maintenance Department to modify its working plans accordingly. Long-range aircraft will receive an individual correction factor, increasing their fuel burn in a certain percentage, while short/medium-range aircraft manual will take a single factor for all the aircraft of the same type or with the same specification. A general scheme of this process is in Fig. 7.3.

It is interesting to dissociate the airframe and systems performance degradation from the engine deterioration, because engines can be the subject of individualized analysis and replaced if it is considered as convenient. Most airlines have ECTM (Engine Condition Trend Monitoring) programs that follow the evolution of different engine parameters, using flight hours and cycles as parameters. The comparison between APM and ECTM provides an idea of what part of the performance degradation is due to the airframe and systems. Figs. 7.4 and 7.5 provide an idea of this concept application for a twin jet. The engines contribute approximately 70% to the aircraft performance.

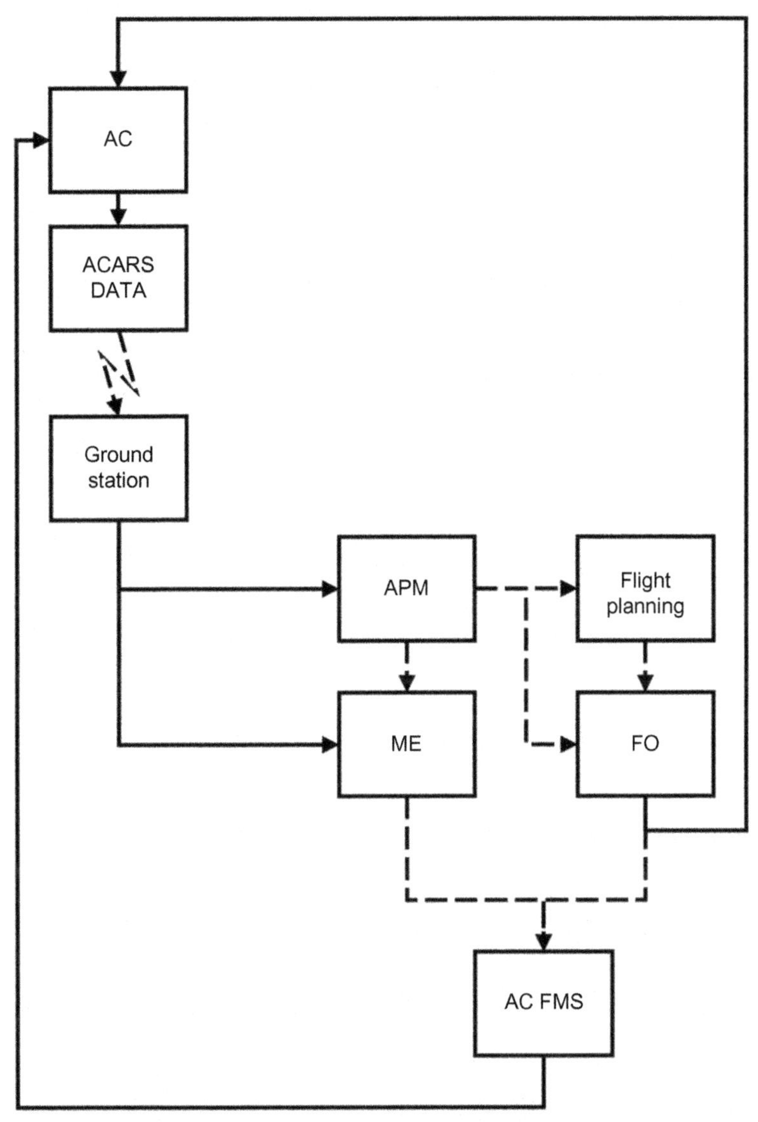

Fig. 7.3 Block diagram of the aircraft performance updating system (*IATA Training: Fuel efficiency and conservation*). AC, aircraft; APM, airplane performance monitoring; ME, maintenance & engineering; FO, flight operations; FMS, flight management system.

Maintenance information can be used for the introduction of immediate repairs in order to recover the lost performance and/or for preventive maintenance, extrapolating the trending lines built upon the recorded data. Even for aircraft of the same type, age, and specification, their performance evolution may differ a lot due to factors related with the assembly of the aircraft, the type of performed missions, and a number of unexpected external factors. Fig. 7.6 shows the evolution of different aircraft of the same model.

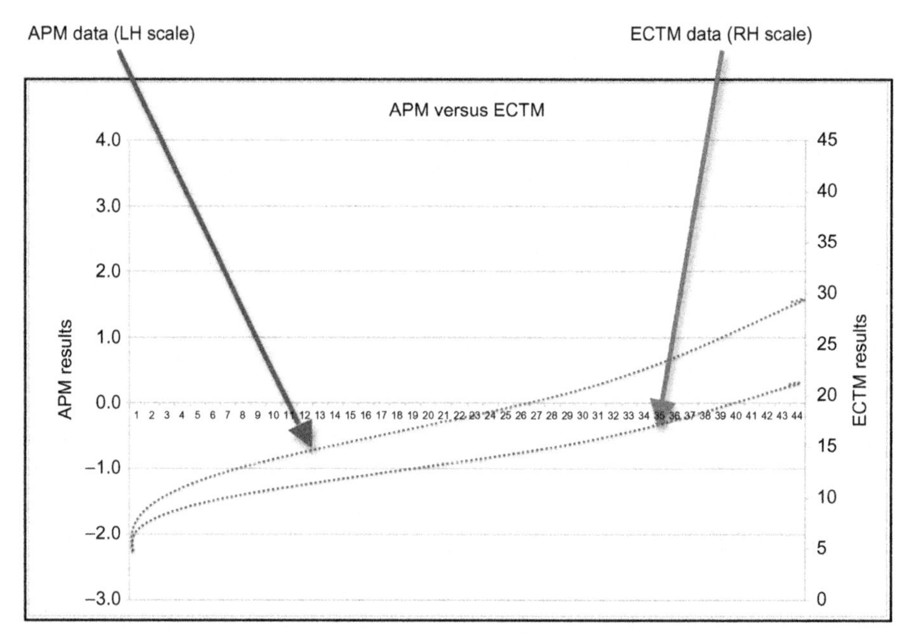

Fig. 7.4 Performance monitoring of the full aircraft (APM) versus engines performance monitoring (ECTM) (*IATA Training: Fuel efficiency and conservation*).

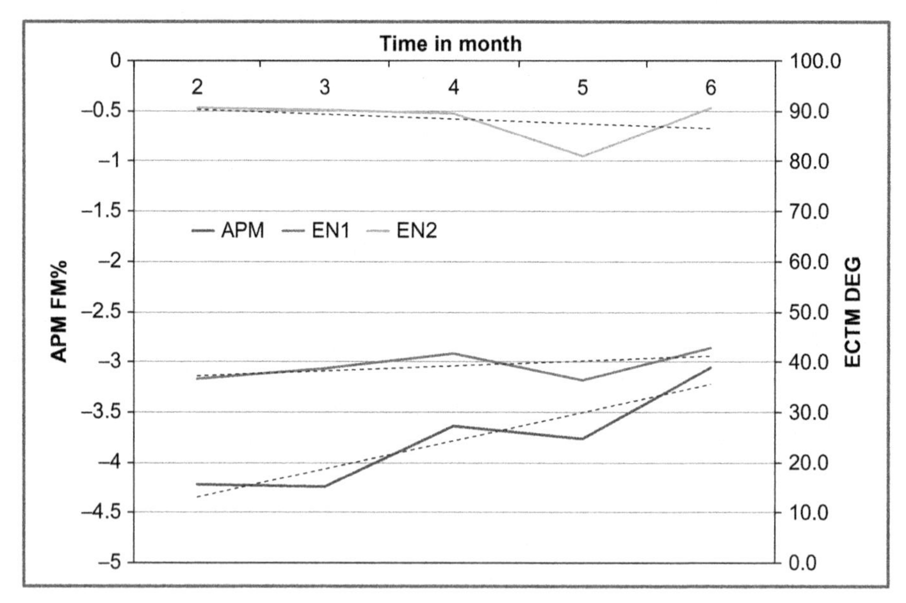

Fig. 7.5 Comparison of the evolution of the twin jet APM versus its two engines (*IATA Training: Fuel efficiency and conservation*).

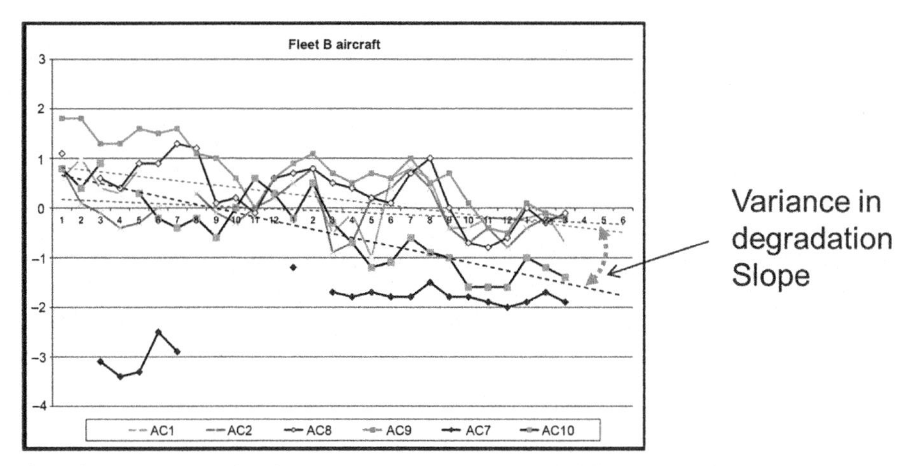

Fig. 7.6 Comparison of performance degradation levels of different aircraft within the same fleet type (*IATA Training: Fuel efficiency and conservation*).

A good knowledge of individual aircraft performance level may be extremely useful for optimizing flight plans, as it was explained in Chapter 5, because it improves the accuracy of the trip fuel calculations and allows to minimize fuel reserves without reducing safety margins.

The main lines of typical maintenance actions with the fuel efficiency increasing purpose can be divided into four categories:

- Engine refurbishment to improve the specific consumption
- Keeping of aerodynamic cleanness
- Reduction of aircraft weight
- System calibration for accuracy optimization

As maintenance tasks referring to different aircraft areas are classified according to the American Air Transport Association ATA 100 methodology (the most commonly used way of structuring the commercial aircraft specification), some of the ATA chapters should receive particular attention for energetic efficiency purposes:

ATA 8. *Weight and Balance*: The OEW is the result of taking the Manufacturers Empty Weight (MEW) and adding up all the items the airline considers adequate to be ready for the commercial service, like the crew and their baggage, passenger service elements and Unit Load Devices (ULDs). A detailed control of their weight may help to achieve a lighter aircraft and save fuel. Related to that is the dispatch CG position. A backwards CG, always within the stability limits, reduces the negative lift of the tail, needed to compensate the nose down momentum created by aircraft weight and wing lift, and gets drag down, resulting on an improvement of the energetic efficiency of the flight.

ATA 21. *Air-conditioning system*: System losses demand more engine bleed and more fuel consumption. Leak tightness should be checked in regular periods or when a decrease in performance is noted. The performance of the individual packs needs to be periodically monitored as well.

ATA 27. *Flight controls*: A correct flight control surfaces rigging reduces the shape drag of the aircraft and improves energetic efficiency. Moveable surfaces like flaps, slats, ailerons,

spoilers, or tabs may suffer mechanical deformations and not fit exactly in the expected position. Calibration of no-deflection position is the most common procedure to minimize this problem. The possible consequences are drag increase due to loss of the skin surface continuity and due to the boundary layer perturbation when appears a gap between the upper and lower part of the lifting surfaces, either wing or stabilizers.
ATA 28. *Fuel system*: System losses are relevant here because they may be in form of fuel spill, a safety issue, or adopt the shape of wrong quantity measurements. The system is volumetric and its accuracy depends on the sensor precision and on the geometry of the tanks. A deformation of the tanks may induce erroneous indications of the fuel inside them.
ATA 31. *Instruments*: Indication accuracy of the main flight parameters (speed, altitude, fuel onboard) is relevant for the fuel consumption as the trip is done out of the flight plan optimum calculation inserted in the FMS, besides becoming the source of safety incidents.
ATA 36. *Pneumatic system*: Same as in the case of the Air-conditioning system.
ATA 49. *APU*: Although the APU running time used to be modest, in short-medium range operation it can be in use up to 1.5 h per flight, depending on the turnaround time and the specific airport equipment. Periodic performance checking would save sizeable amounts of fuel.
ATA 51–57. *Structure and doors*: The weight and the condition of the structure have two effects on the energy efficiency. Along their useful service, life aircraft are gaining weight due to modifications, repairs, dirtiness, humidity, and other miscellaneous events. At the same time, deformations caused by hits or repairs may increase drag and the poor adjustment of doors (both passenger, cargo or service doors) collaborates to that drag increase. The first effect is discussed in Section 7.4, while aerodynamics is in Section 7.3.
ATA 72. *Engines*: The most important element for energetic efficiency is obviously the power plant that typically requires 50% of the total maintenance cost. Their treatment is detailed in Section 7.2.

Deviations from the aircraft specification may be allowed under a number of conditions in order to ensure adequate safety levels. Commercial aircraft manufacturers provide a Manufacturer Minimum Equipment List (MMEL), approved by the aircraft designer State CAA, and operators use it to elaborate their own Minimum Equipment List, adding those elements essential for their type of operation. The MEL lists the conditions under which the aircraft may be dispatched with specific equipment inoperative. The conditions include the period during which the aircraft can be operated with the system inoperative (typically 10 days plus some possible extension in special conditions) and, in some cases, requirements for additional fuel load. The most common cases demanding more fuel are system elements failure that needs other systems actuation (AC generator failure replaced by additional APU running time), or the failure of a valve or moveable element in an open position, wasting some energy (an anti-ice valve or a foldable landing light).

The configuration deviation list (CDL) is a part of the AFM that allows the aircraft to be dispatched with specified components not fitted. All components must be reinstalled at the earliest maintenance opportunity (around 1 week, according to the airline State CAA authorization). For items, whose loss or failure will bring a fuel consumption penalty, it is beneficial to make special efforts to replace them as soon as possible. Airlines used to have a synoptic table indicating MEL/CDL systems and components, the limitations imposed by their condition, and their impact on fuel consumption. That table gives a clue of the fuel cost of not having performed the corresponding maintenance action. Most common cases involve some external structural parts, like track fairings or tip fences.

7.3 Engine refurbishment

The APM work is based on the whole aircraft performance, but the most important elements in terms of fuel efficiency are the engines, generally responsible for 70%–75% of the total results. Engines can be replaced several times along the operative life of an aircraft by repair need, feature upgrading, excessive performance degradation, or logistic reasons.

The engines have a strong initial performance degradation during the first 30–60 cycles that is not recovered. It is the result of the adjustments among the different moveable and static parts of the engine, working for the first times at a wide range of temperatures and mechanical efforts. After this gap, the increase in fuel consumption follows a continuous way. About 80% of this degradation by use can be recovered in a major overhaul. As the maximum timing of maintenance actions is dictated by the LLPs (Life Limited Parts), the convenience of more frequent repairs, addressed to performance resaturation, depends on the tradeoff between repair cost and fuel savings.

Typical indicators of engine performance are exhaust gas temperature (EGT) margin and specific fuel consumption (SFC). The first one is the temperature of the gases when leaving the engine turbine and provides the temperature limit for operating it. As the engine working efficiency decreases with time, more fuel is needed to reach the required thrust and EGT increases. Cockpit indicators include an EGT "red line" not to be surpassed and the EGT margin is the difference between actual temperature and the red line one. When margin disappears, engine thrust has to be reduced or the engine must be removed for maintenance and EGT restoration.

SFC measures the engine fuel consumption per unit of thrust and increases as engine efficiency falls, having a direct effect on the mission fuel of the aircraft, but not affecting safety unless the associated EGT increase goes up to the red line. Both indicators are tracked in the ETM and are the main parameters to be analyzed at the moment of planning the maintenance actions and decide the best moment to remove engine and perform an overhaul. The extent and the cost of that overhaul and refurbishment have to be traded off against the level of EGT and SFC recovery.

In the routine engine service life, the most important sources of engine deterioration (see Fig. 7.7) are:

- Dirtiness accumulation
- Leakages caused by seal degradation
- Increasing blade tip clearances
- Deterioration of the blade aerodynamic profile due to small dents and erosion

The last three problems may need major maintenance actions to be solved, but the first one can be treated by periodic compressor washing, with the engine in the shop or under the wing.

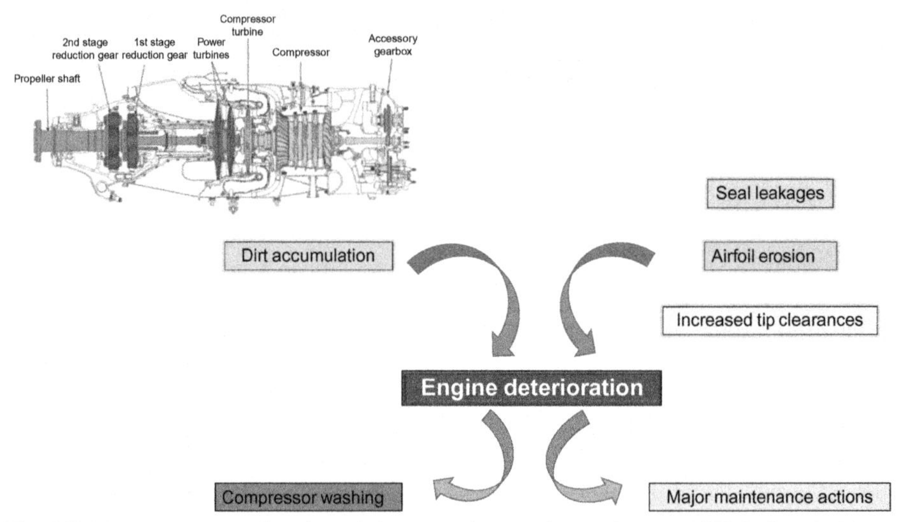

Fig. 7.7 Most common engine degradation areas in a modern turboprop (*ATR fuel saving*).

Dirtiness is absorbed through the fan flow, most of it during the high thrust maneuvers close to the ground (takeoff run and landing with reverse use). A part of the entering powder and foreign objects get out through the secondary engine flow, but some are ingested into the low-pressure compressor after passing by the fan, causing damages to the blades or getting stacked to them, changing aerodynamic profiles and reducing efficiency.

A periodical compressor wash with pressure water may recover up to 1.5% of SFC (specific fuel consumption) per operation if it is implemented at the proper interval, because the cost of aircraft immobilization and the washing operation need to be compensated by the fuel savings. Compressor washing is not included in the aircraft maintenance manual and each airline can decide the best way to perform it, selecting the intervals according with the level of degradation recorded by the ETM. Periodic engine washing also helps to keep the EGT margin stable and, therefore, widen engine overhaul intervals.

Fig. 7.8 shows the effect of compressor washing, comparing the performance degradation in the no-wash case versus two different washing sequences, every 500 and every 1000 cycles.

Fig. 7.9 gives an example of the degradation and the washing effect in different engines of similar age, measure in terms of FF (Fuel flow) and EPR (Engine Pressure ratio). The lines go up (more degradation) and move down when a compressor washing is made. It is interesting to notice that A340/600 is a long-range aircraft with a large average stage length. Then, 500 cycles may mean 1 year in service.

The levels of contamination are very variable, depending on the type of operation and the kind of environment in which the aircraft moves. Some manufacturers advise to start with one washing a year and adapt the frequency of later actions to the performance trend monitored by the ETM system.

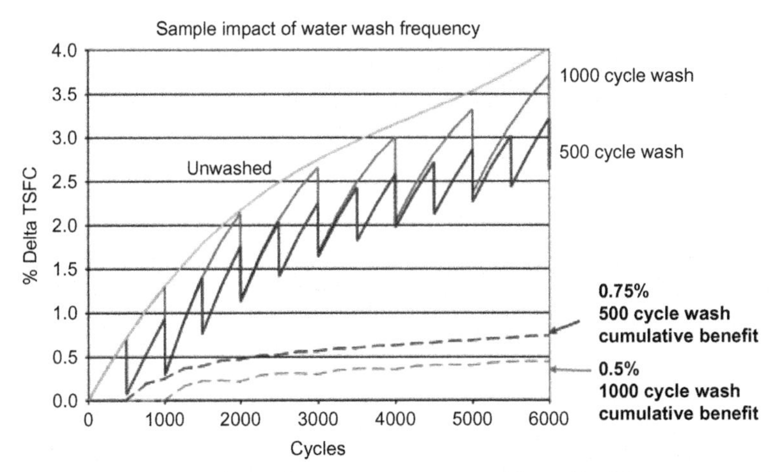

Fig. 7.8 Schematic graphic of the effect of compressor washing with different frequencies (*IATA Training: Fuel efficiency and conservation*).

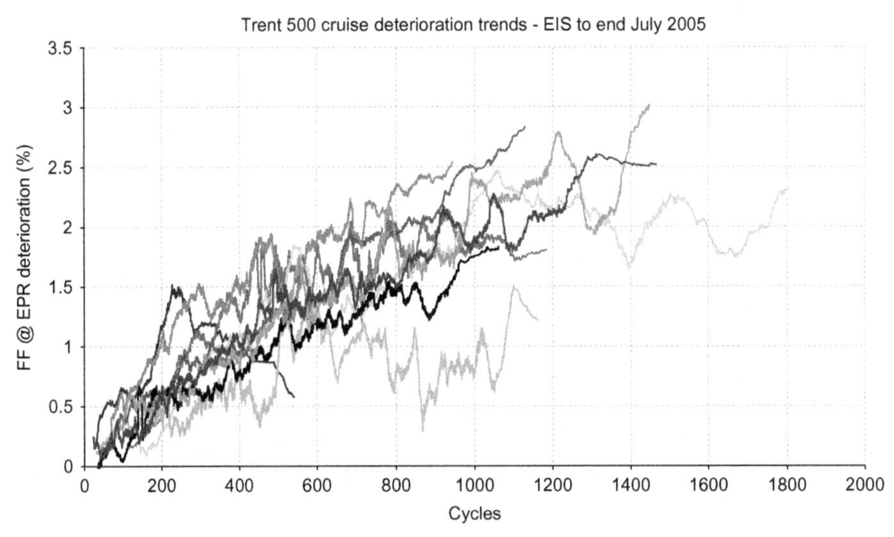

Fig. 7.9 Comparison of Fuel flow/EPR deterioration in different Rolls&Royce Trent 500 engines in the Airbus A340/600 (*Benito, A.: Technology and maintenance for fuel saving*).

The standard washing procedure is based on water and cannot be applied at ambient temperature below 5° centigrade due to the possibility of freezing. An interesting alternative has been developed by Lufthansa Technik and the University of Darmstadt, also injecting a cleaning agent into the engine, but replacing the water by CO_2 dry ice pellets. A compressed air system injects the three by 6 mm pellets into the engine. When they hit the blades of the compressor stages, the kinetic energy released by the impact loosens some of the dirt. The dry ice pellets then turn into gaseous stage and no residue remains in the engine. As the dry ice temperature is around −80°C, the dirt contracts and becomes brittle, being easily removed without leaving any waste product behind.

Seal degradation is one of the most common events that appear with engine aging, particularly in hot areas. It makes rotating assemblies lose efficiency and increase engine temperature to obtain the same thrust, consuming more fuel. ECTM is a good tool to detect this phenomenon that can be localized through boroscopic inspections. Seal repair requires modular disassembly and a high amount of working hours.

Other similar problem, but this time in the cold area, is posed by the deterioration of the seals around the thrust reversers that may disturb the clean passage of the secondary flow air and create some overboard leakage.

The aerodynamic efficiency of engine rotating elements is strongly dependent on keeping blade tip clearances at their design limits. Increasing clearances reduce pressure ratio and efficiency in both compressors and turbine modules. On the contrary, too narrow clearances increase the probability of friction between the tip of the blade and the casing with a minimum level of ovalization or a little bending of the engine shafts. The result is damage in the blades that may lead to loss of efficiency and, in the worst case, to the failure of the blade with the corresponding engine damage. Again, a combination of ECTM and boroscopic inspections should detect the problem at its origin and avoid major damage.

In a similar way, foreign object ingestion often causes damage on the rotating elements of the engine. The ones most likely to be reached are the front parts, like fan or low-pressure blades, often around the leading edge zone. In addition to the already-mentioned ECTM and boroscope combination, visual inspection is a great help in detecting the most important damages.

7.4 Aerodynamic cleanness

Commercial aircraft are complex mechanisms integrated by a number of rigid and moveable parts. The optimum efficiency is achieved creating the lowest amount of drag in each one of the operating configurations that requires the airframe to be free from any irregularities. Surfaces should be as smooth as possible, doors and panels should be flush with surrounding structure, and all control surfaces should be rigged to their specified positions.

With the service incidences, the surface of the aircraft suffers a number of small hits; it is eroded by hard elements friction and gets dirty. The result of all those irregularities is an increase of drag and, in consequence, additional fuel consumption. The most damaging effects are located in the front sections of the aircraft, nose, cockpit, wings, and tail leading edges as indicated in Fig. 7.10.

Other elements that may increase drag are movable parts losing their adjustment with the fixed structures they are attached to. Flight controls, as ailerons, spoilers, flaps, rudder, and elevators, need to be periodically rigged to adjust their neutral position and achieve a smooth continuity with the wing or tail surfaces. It is important

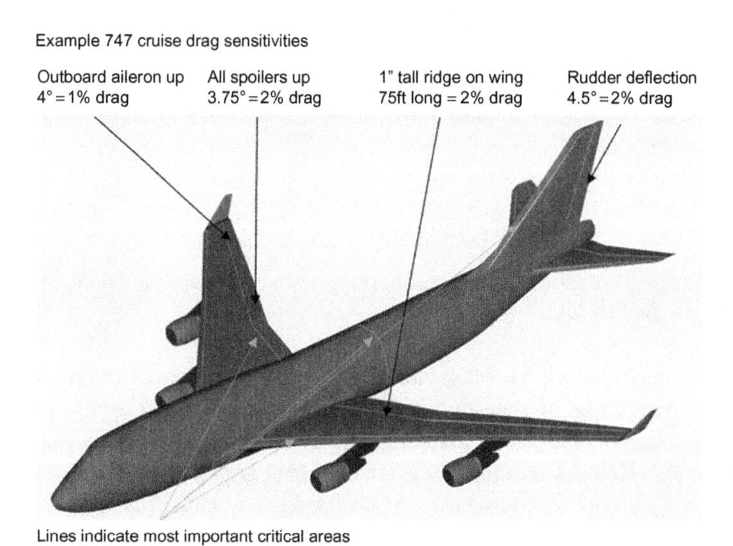

Fig. 7.10 Critical areas for aerodynamic drag creation (*IATA Training: Fuel efficiency and conservation*).

to take special care of the spoilers, because they are deployed in short periods and their position in the upper wing surface where little irregularities have a great impact in drag. These irregularities happen when a spoiler is not flush with the wing profile creating turbulence and increasing drag. The total drag increase by all these reasons may get up to 0.4%.

In some cases, the misalignments break the smoothness of the surfaces but some of the moveable elements are designed to open a gap between the fixed part they are attached to and themselves during certain phases of the flight, like flaps in approach or in landing. If these gaps get too wide or do not disappear in the retracted position boundary layer over the upper surface is blown up increasing turbulence and aerodynamic drag. Fig. 7.11 shows a schematic description of this situation. Periodical rigging is needed to ensure a correct functioning of the moveable surfaces.

The importance of all these problems in terms of fuel efficiency varies a lot depending on the type of aircraft and the range of the flights. Drag increase effect increases with speed and duration, becoming larger at very long flights at high-speed cruise regime. In the movable control parts, leading edge slats and wing spoilers rigging are the key elements of maintenance action. As an example, a 10-mm spoiler misrig in an A320 would increase 21 kg of fuel consumption in a 3-h flight.

Some areas of the aircraft surface use seals to ensure the continuity between two rigid or moveable surfaces. In some aircraft, the transition between wing and fuselage in made through fairing panels, not a part of the primary structure. Flexible seals, known as Karman seals, ensure the smooth airflow along the converging structures of wing and fuselage and request careful maintenance to keep their geometric shape for minimizing drag. Flexible seals are also used to fill external gaps between moving surfaces and access panels and their surrounding structure.

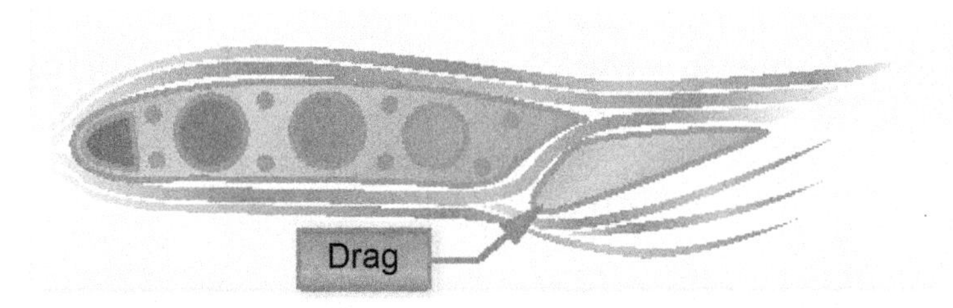

Fig. 7.11 The effect of lack of sealing in a moveable surface (*Benito, A.: Technology and maintenance for fuel saving*).

A third aircraft surface problem may be the misalignment of doors and registry panels, increasing drag, or even diminishing cabin airtightness. Larger doors include passenger and cargo, emergency exits, landing gear doors, particularly the nose landing gear, and service doors. Other smaller elements are APU doors, avionics panel, moving cockpit windows or hatchets, and all the registry panels used in checking and maintenance operations.

Poor-quality repairs are having similar effects, when they break the continuity of the surface increasing roughness by nonaligned elements or leaving gaps between two structural parts that should be in contact. Fig. 7.12 shows an example of this situation.

The seals of the cabin doors have a dual function. On one hand, they close the gap between the door and the surrounding structure, but also have to avoid any air leak due to the pressure difference inside and outside the cabin. Cabin air leaks are important for two reasons: they mean a loss of energy to be covered by additional fuel consumption, and they have a great effect on drag, because escape in a direction perpendicular to the fuselage skin, altering the local airflow and creating additional drag.

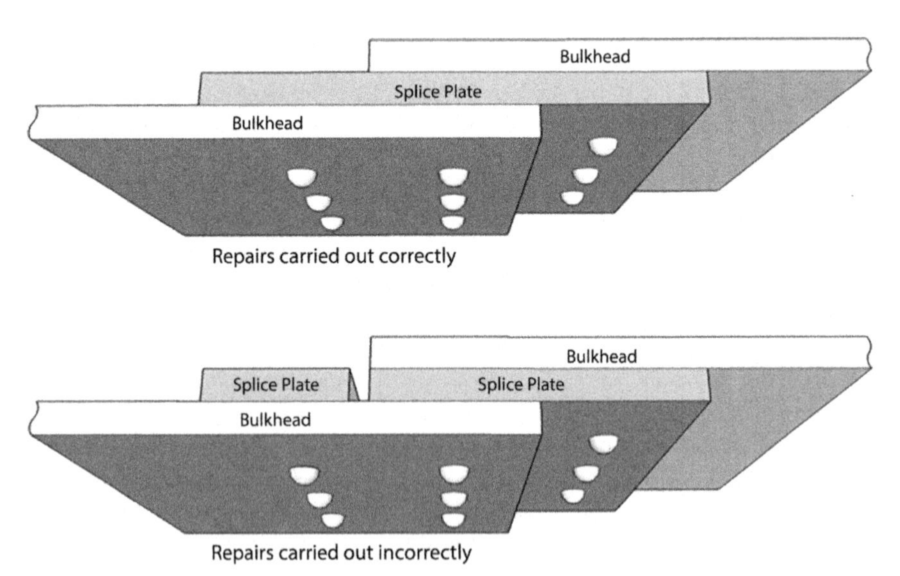

Fig. 7.12 Irregularities from a poor quality repair (*ATR Fuel saving*).

Finally, the surface of the aircraft gets dirty during the normal operations and the overall paint layer deteriorates. Both events are more or less frequent depending on the operating environment of the aircraft. The thickness of paint used to be small, between 1/2 and 1/3 mm, but when it is lost in critical areas (nose, cockpit, wing surfaces) local airflow is upset and overall drag increased.

The natural accumulation of dirt on the external part of the aircraft creates slight roughness, inducing additional drag as well. Aircraft operating in windy places, with many sand grains in the air are more likely to accumulate small material parts on the skin and will need more frequent cleaning. In some cases, aircraft washing is made attending more to airline image considerations than to fuel efficiency target.

A standard wide-body aircraft is washed three or four times a year, consuming about 12,000 L of water that gets contaminated with cleansing agents and dirt and needs to be filtered. The process may take between 8 and 10 h to complete. An alternative procedure, known as dry wash, consists in applying a thin layer of cleaning paste to the aircraft surface, after having protected sensors and landing gear. Fig. 7.13 shows a cleaning crew in action with the dry wash procedure.

This paste bonds with dirt, and after polishing and drying, it seals the surface. This procedure takes longer (about 20–24 h) and requires more man-hours, but it is more effective and saves the task of purifying the water used in the wet wash process. It is not applicable to the landing gear because chemicals are not compatible with brakes and tires, which needs to be washed with water.

Washing and polishing the aircraft surface may reduce drag in the order of 0.1–0.2%. It is worthwhile to remember that aircraft washing is also important from the public relations policy as a clean, bright airplane has a much better view.

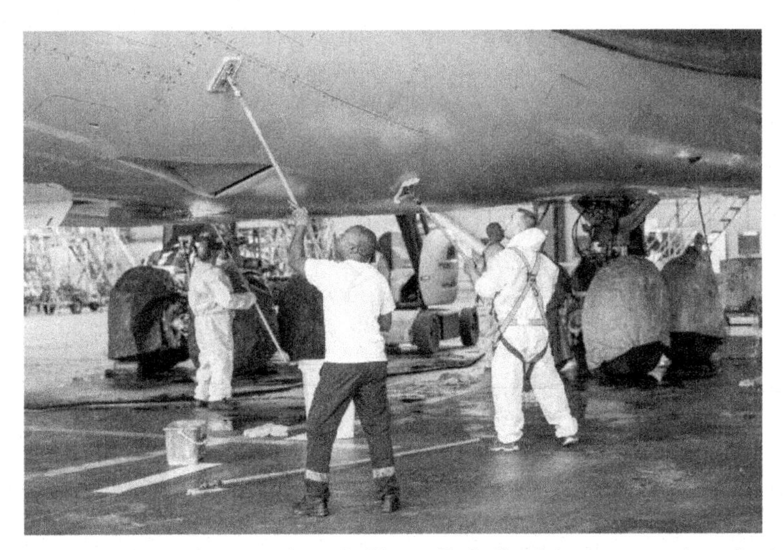

Fig. 7.13 Dry-wash-cleaning procedure (*Lufthansa Technik: Maintenance and overhaul technologies*).

7.5 Reduction of aircraft weight

The aircraft OEW increases about 1% in a period of between 5 and 10 years and should be controlled systematically. Airlines do not weigh commercial aircraft very often, in most cases after heavy maintenance actions like Intermediate Layovers or Overhaul. Autonomous weighing systems installed in the aircraft undercarriage have not proved to be very reliable. Then, OEW used to be corrected and new values going in to the Flight Manual once every several years. Like in the case of the specific fuel consumption, long-range aircraft may have an OEW per tail number due to the possible payload penalties, while short/medium-range fleet used to have aircraft-type average values.

A nonexhaustive list of weight gains is:

- Dirtiness and humidity accumulation in nonaccessible areas
- Additional weight as a consequence of repairs
- Service Bulletins and other embodied modifications
- Empty ULDs
- Unneeded emergency or maintenance equipment

There are some areas inside an aircraft not easily accessible in routine maintenance and only open during heavy structural maintenance, like D checks. That check offers the opportunity for removing dirtiness out of hidden places and drying the zones where humidity is accumulated, like fuselage isolation blankets. Condensation is caused by the high altitudes and low temperatures, and is greater for high utilization aircraft that spend more time in those conditions. Very short turnaround times may also reduce the drying periods. A middle-size aircraft, like a B737 or an A320, can accumulate and retain about 200kg of water, which is roughly translated into seven additional fuel kilograms per flight hour.

Maintenance actions asking for replacement of a determined element or component are not adding any weight to the aircraft but those intending to repair a damaged part use to add some weight. In many cases, repairing is cheaper than replacing but the possible weight increase must be taken into account when calculating the economic repercussions of the decision.

Along the life of the aircraft, its specification is modified by the addition of new systems, the modification of the existing ones, or the replacement of the originals by new equipment, with some weight effects.

A very common example is the change of the cabin interior configuration to support a new commercial service strategy, something that is going to happen several times during the service life of an aircraft. The interior change in a large long-range aircraft may add several tons to the OEW.

The economic evaluation of the planned modifications is generally made on base to time needed to recover the investment that used to be between 2 and 5 years. Fuel burn changes must be taken into account for that calculation.

In addition to calculate carefully the weight repercussion of the changes, special attention should be paid to remove all the elements becoming useless because of the modifications. Large commercial aircraft often have installed some provisions, as the designers take care of future specification upgrading, expected new regulations or additional uses of the model. Space provisions provide just a place to install the expected new equipment, but structural provisions or wiring provisions add physical elements with additional weight. If those provisions are not going to be in use, they should be removed to save weight.

All wide-body and some standard body aircraft use standard ULDs to carry freight and passenger baggage. Most of the airlines include the weight of these pallets and containers in their OEW calculations. However, in some cases, they are flown empty as there is no cargo for them in a specific flight and they become a useless weight, burning additional fuel. In addition, baggage containers may go almost empty as handling personnel try to separate baggage to destination from baggage making connections to different airports. This practice facilitates turnaround operations in shorter time, but an economic analysis is required to ascertain the real cost of it.

Aircraft may carry emergency kits, fly away kits, spares or other maintenance and support equipment, uploaded for a specific operation in a certain period and left there afterwards. Close monitoring should be done to avoid forgotten items to continue flying when they are not needed anymore. Regular cleaning of belly compartments is a good practice because foreign objects, like suitcases parts (handle, wheels), add weight and can damage cargo compartment panels.

A useful but more complicated practice is to adapt aircraft equipment to the specific conditions of each flight. Very short-haul operations may not need certain galley equipment as water heaters, coffee makers, or ovens that can be temporarily removed. One additional step forward would bring us to replace slide rafts, needed to comply the floatability regulations on over-water missions by basic slides when the flight is over solid ground. Weight saving is about 30 kg per unit.

A very effective weight reduction procedure is the monitoring of the potable water uploading. The majority of airlines are providing bottled drinking water to the passengers and the liquid inside the potable water tank is used just for the washing hands and flushing the lavatory in the toilets. Depending on the number of passengers and the length of the flight, the amount of water can be reduced. An A320 tank can take 200 kg of water and one toilet flush is about 0.2 kg, then in a short flight, half the tank capacity would be more than sufficient. Airlines make a double-entry table with the aircraft seating and the length of the flight to calculate the potable water needs.

While some aircraft types have a refilling system with a level preselection scale allowing doing that in a very automatic way, others have no other indication than empty or full. There is the possibility of calculating the amount of water unloaded by measuring the replenishment time, if the flow rate is known. In a similar way, but in a different direction, toilet waste is stored in a waste tank. Emptying that tank as often as possible saves an additional amount of weight.

Although it is not exactly a maintenance item, when discussing weight reduction there is always a hectic discussion on the weight-saving potential of on-board service provisions. Quantity of food, beverage, duty free, pillows, blankets, newspapers can be dimensioned in one or other direction and their weight reduced by specific actions, like using plastic bottles instead of glass ones or reducing the number of on-board company magazines, or even edit them in a lower-quality lighter paper. Some marketing policies determine some practices, like taking round-trip catering, that move in the opposite direction than saving weight.

7.6 System calibration

Flight systems offer information within a predetermined level of error margins. A larger-than-specified error may induce the crew to operate the aircraft in less-than-optimum conditions in terms of speed, altitude, or even center of gravity. The magnitude of the penalties can be very high, as the flight will not be done according to the flight plan, probably optimized with the use of a Cost Index. A Boeing B747-400 flying 0.01 Mach faster than expected would increase fuel consumption by more than 1%.

The basic air data needed for a correct navigation are true airspeed, pressure altitude, ambient air temperature, angles of attack and sideslip, rate of climb, and Mach number. The pressure signals collected by sensors are converted mechanically into indications on the altimeter, vertical speed and air speed indicators, and Machmeter. In a similar way, outside-air temperature and local air flow angles send signal to the cockpit instrumentations for sideslip and attack angles and temperature. The accuracy needed for a particular application indicates the best way to measure air data and the amount of calibration effort required.

In terms of fuel consumption, speed and altitude are the most influential parameters. The calibration of on-board instruments is difficult and in most of the cases, the instrument itself is calibrated in the test bench. The sensors calibration can be done by comparison with an item experimented inside a wind tunnel facility. Not every airline can do these types of things by itself and the most common practice is to replace the instrument when an anomaly is detected and send it to the manufacturer for repair.

The air data system of some aircraft models does not allow calibration and any system failure results in a crew warning of malfunction. Even in those cases, regular cleaning of the air data system sensors and the corresponding network of piping for processing the air is in the maintenance program but can increase its frequency if found beneficial.

Infrastructures

8.1 Introduction

Air Transport Infrastructure integrates all the ground facilities needed to support airline services with the adequate levels of safety, reliability, and economy. The two main elements of those facilities are airports and air navigation services.

From the management point of view, they are very different. Airports are ground infrastructures, placed generally close to the cities they serve, acting as interchange center of different transport modes and including many commercial services that take advantage of the great number of passengers going in and out through the terminals. In terms of energetic efficiency, the weight of pure aeronautical services may be small, compared with the energy consumption of the other activities, in particular in the case of large airports. They need to be certified by their own State CAA. While most of the airports management is subject of State regulation and control, many of them have a private, business-minded management, are quoted in the stock market, and are able to compete in an open entrepreneurial environment.

Air navigation services are, with a small number of exceptions, either part of the State administration departments or corporatized state-owned companies, with modest private enterprise participation. ICAO divides world airspace in areas under the control of ATM facilities of individual States, with neither overlapping nor empty zones and efforts are done to improve coordination and collaboration rather than competition.

In both cases, the amount of energy consumed in providing aeronautical services is relatively small compared with aircraft flight consumption. The fact of developing their physical activities on the ground makes possible the use of alternative energy sources and reduces their environmental footprint to the extent of becoming carbon neutral, a target achieved by some airports in recent years. A majority of the programs applied by individual airports have three simultaneous goals: reduce the energy bill, reduce energy consumption, and reduce its carbon footprint. In most cases, the three goals are aligned but, in others, an airport may decide to increment its energy use if an alternative source provides cheaper energy than the one having the highest efficiency.

For the airports, a key factor is to decide which elements and to which extent they belong to the energy efficiency calculation area. For example, ground traffic arriving at the terminals may or may not be included in the energy calculations. If the answer were positive, next step would be to decide the distance around the airport to be calculated. Same type of questions appears when analyzing the energy consumption of the different buildings in the area, not strictly devoted to aeronautical functions, like hotels, warehouses, shops, car rentals, and convention centers.

In addition of their own energy efficiency magnitudes, air transport infrastructure collaborates to optimize the performance of the flights, trying to adapt flight itineraries to the operating aircraft and routes requirements, and minimizing delays both in flight and on ground. Intergovernmental Panel for Climate Change (IPCC) experts consider that the fuel-saving potential of these activities may reach the 12% figure.

Energy Efficiency in Air Transportation. https://doi.org/10.1016/B978-0-12-812581-6.00008-9

8.2 Air navigation services

The structure and operation of the airspace is an exclusive competence of each individual State, complying basic technical rules established by ICAO, and is usually limited by national borders, including the 12 nautical miles area of coastal seawaters. When required, a national Air Navigation Service Provider (ANSP) facilitates ATM beyond national borders over part of high sea (according to an ICAO's agreement) or in part of an adjacent State's airspace, when an airport is very close to the border and it has a difficult access from its own State territory.

There are also a few cases when there is a multinational ANSP, as ASECNA (*L'Agence pour la Sécurité de la Navegation Aérienne en Afrique et à Madagascar,* the agency for aerial navigation safety in Africa and Madagascar), COCESNA (*Corporación Centroamericana de Servicios de Navegación Aérea,* the Central American Agency for Air Navigation Services) or EUROCONTROL Maastricht Control Center in Europe, providing air navigation in more than one State.

Each National Authority can decide which parts of its airspace are available for civil aviation use and in which conditions. Part of the airspace may be reserved for military use, or simply prohibited to fly as it is over military bases, strategic facilities, governmental dependencies, or other places that the State is interested into protect and leaving free of undesired interferences.

Optimization of flight tracks with the target of getting as close as possible to the orthodromic trajectory (direct point to point or "as the crow flies") can greatly increase fuel efficiency by reducing the flown distance. However, this is not the only factor to be taken into account. Global airspace capacity is a key issue because it can facilitate the reduction of congestion, eliminating holdings, diversions, and other disruptions of the flight plan and, at the same time, ample capacity offer provides operators with more possibilities of optimizing flight trajectory in the four dimensions (three geometric plus time). If airspace is not restricted, each flight can decide the optimum trajectory at each moment of time, considering aircraft features (weight, thrust, speed) and external variables, like actual position, temperature, wind, or altitude.

A great part of the effort by the aeronautical community in the last years has been devoted to open new routes, optimize existing ones, and develop a better coordination among CAAs, ANSPs, airports, airlines, and military authorities who, in many cases, are the owners of airspace zones of dual civil and military use or making frontier with the civil use areas.

Civil military cooperation is a good example of the difficulties of airspace refurbishment. Chicago Convention determines that every civil flight has the right of flying inside the airspace of any State having ratified that Convention. Notwithstanding the States have the right of blocking airspace areas due to specific reasons (security, military use, environmental protection, etc.), with the condition of not discriminating between foreign and national operators.

The amount of airspace reserved for those uses changes from country to country, but it may rise up to 40% of the national total. Civil operators have to avoid military zones, a limitation added to the necessary separation between the two types of flight operations: Visual Flight Rules (VFR) and Instrument Flight Rules (IFR), making difficult the trajectory optimization.

Some of the restrictions are permanent, like the overflown of military bases, governmental buildings, or sensitive industrial facilities as nuclear power plants. Others cover a limited altitude only, as the flying height limits over natural parks and wildlife reserves. There are also some temporary prohibitions such as the zones for military aviation exercises. When all the limitations concur, the difference between the orthodromic trajectory and the real one can be as high as 15%–20% with the corresponding increase in fuel consumption. Fig. 8.1 illustrates a real case of a flight between Amsterdam and Rome. On that day, permanent restricted areas were joined by other temporary restrictions due to military exercises.

Civil-military cooperation used to be smooth (EUROCONTROL, for instance, has a permanent coordination committee with representatives of both groups), but proposed changes to the present status may be very expensive if they require to move military facilities to other place or take further away the area of military exercises. Then, the problem becomes budgetary due to the additional time and fuel consumption military aircraft will suffer for flying longer distances in their training missions.

The respect to the States' sovereignty in their airspace does not preclude the adoption of international measures that need the participation of multiple States, in order to ensemble their frontiers and coordinate the ATM procedures. A number of different procedures have been applied or are in the implementation process to optimize time-related flight sequences and trajectories in the horizontal and vertical planes.

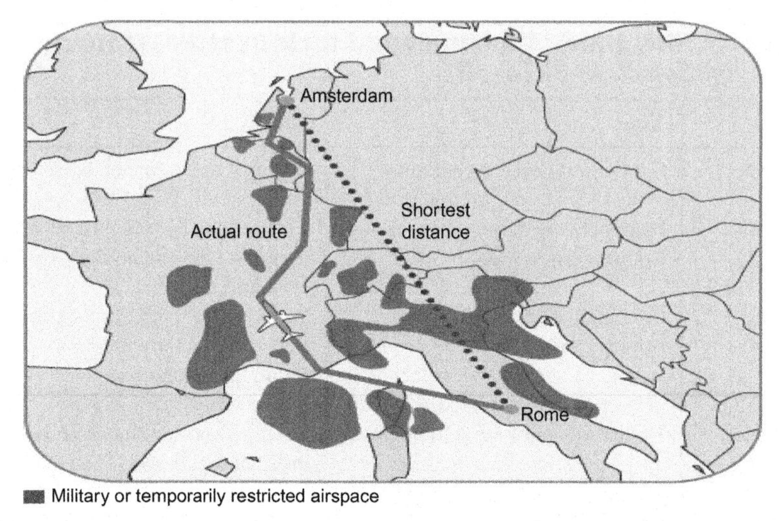

Fig. 8.1 The trajectory of a Rome to Amsterdam flight, avoiding restricted areas (*IATA Training: Aviation and the Environment*).

8.3 ICAO global air navigation plan

The improvement in the ATM efficiency has two main elements: development of new and advanced technologies and the introductory timeline of those elements in the world different flight regions. New technologies can be tested on ground and in some airports equipped for that purpose, like EUROCONTROL's Bretigny facilities, but once their introduction is decided, the deadlines become very long, because all users (ATC, airports and airlines) have to progressively modify their equipment and operational procedures in order to adapt to the new situation.

During the 2011–13 period, ICAO has agreed a GANP to organize the zonal introduction of those elements into a time-flexible series of performance improvements and timelines. The pace of the process is determined by individual States, according to their available technology and resource levels. This resolution was supported unanimously by aviation stakeholders and was structured in four Aviation System Block Upgrades (ASBUs), with the following temporal sequence:

Block 0	Year 2013
Block 1	Year 2019
Block 2	Year 2015
Block 3	Year 2031 and onwards

Each one of the Blocks was divided into different modules. Table 8.1 shows, as an example, the already-implemented Block 0 modules and the type of the most important expected benefit of each one.

Table 8.1 ASBU Block 0 modules and their benefits ***(ICAO: Global Air navigation Plan 2016–30)***

Module	Title	Benefits
B0-CDO	Continuous descent operations	Reduced fuel burn and noise on arrival
B0-FRTO	Free route operations	Reduced in-flight fuel burn
B0-RSEQ	Runway sequencing	Reduced airborne holding and taxi-out time
B0-CCO	Continuous climb operations	Reduced fuel burn during climb
B0-NOPS	Network operations	Reduced fuel burn in all phases of flight, including taxi
B0-TBO	Trajectory-based operations	Reduced in-flight fuel burn
B0-WAKE	Wake turbulence separation	Reduced taxi-out time and reduced in-flight fuel burn
B0-ACDM	Airport collaborative decision making	Reduced taxi-out time
B0-ASUR	Alternative surveillance	Reduced in flight fuel burn
B0-APFL	Optimum flight Levels	Reduced in flight fuel burn

Regarding the energetic efficiency, the main action lines applied in these coordination programs can be divided into six groups:

- *Flight information integration*, with participation of ATM, airports, and airlines, compiling together all the flight data since the initial planning phase. The purpose is to put the conditions for a free of incidences flight without any avoidable delay and, at the same time, be able to monitor the flight development and take the adequate actions when a disruption appears. The first concept is known as Gate-to-Gate plan and needs to be implemented by each one of the stakeholders. Airport, airline, and ATC adjust their schedules in such a way that the aircraft starts engines and do the taxi out, takeoff, and climb without any waiting time or delay. To do so, everybody must be prepared for a real-time interchange of data, with digitized and data-driven Collaborative Decision Making (CDMA). This is complemented by the continuous monitoring of the controlled flights in the airspace, capable of suggesting the most likely solutions to the unexpected incidents. EUROCONTROL system for this purpose is the Central Flow Management Unit (CFMU), placed in Haren, close to Brussels. CFMU is fed with all IFR flight plans in the European airspace and can act as a monitoring system but it is capable to perform simulations and provide results in term of fuel burnt, flight time, and economic repercussions. Suggested solutions for real-time problems are typically optimized in terms of accumulated delays of all the affected flights, a solution that may induce a higher number of affected flights, but with very small delays in each one.
- *Terminal airspace development*, structuring the operations with optimum climb/descent profiles and preparing the use of advanced approach/landing navigation aids. Some details of the most frequent low-fuel procedures, like Continuous Descent Approach (CDA) and Continuous Climb Operations (CCO), have been provided in Chapter 6. In any case, the final decision of using them or not is always in the hands of the local ATC responsible.
- *RNAV routes definition*, applying automated traffic management tools. Area Navigation (RNAV was originally coming from *Random Navigation*) can be defined as a method of navigation that permits aircraft operation on any desired course within the coverage of station-referenced navigation signals, or within the limits of a self-contained system capability, or a combination of these. It requires a determined performance level, defined by the required navigation performance (RNP) category. In accordance with this, performance-based navigation (PBN) specifies that aircraft RNP and RNAV systems performance requirements are defined in terms of the accuracy, integrity, availability, continuity, and functionality required for the proposed operations in the context of a particular airspace, when supported by the appropriate navigation infrastructure. In 2010, ICAO and IATA jointly established a Task Force with the objective to disseminate the best practices of the global and regional structures already put in place for PBN implementation. At this moment, more than two-thirds of the 191 ICAO Member States are committed to implement PBN and are in different stages of the process.
- *Flexible airspace use (FAU)*, building up the capability of flying within forbidden areas (military or reserved zones) in some periods of time or under certain preestablished conditions. This is in principle a coordination and information problem, the benefits of which may be important in flight regions with many of those areas and a heavy traffic demand, like Central Europe.
- *Reduce Vertical Separation Minima (RVSM)*, creating additional flight levels to be used by the operators as a mean of flight optimizing. This has been a long ICAO program, started in 1997. Annex 2 to the Chicago Convention established the separation between westwards and eastwards IFR flight levels in order to reduce collision probability. Up to FL 290 (altitude of 29,000 ft.) the separation was 1000 ft. (i.e., eastwards FL 210, 230, 250, 270, 290; westwards

200, 220, 240, 260, 280), but at higher altitudes, it changes to 2000 ft. (eastwards 290, 330, 370, 410, 450, etc.; westwards FL 310, 350, 390, 430, 470, etc.), taking into consideration the lower precision of the old aircraft instrumentation.

- With the progress of the positioning systems, ICAO started the RVSM program to change the upper space separation to the standard 1000 ft. (westwards FL 300, 320, 340, 360, 380, 400; eastwards 290, 310, 330, 350, 370, 390, 410), creating 6 new flight levels to increase 15% air space capacity. As evident, a number of westwards levels in the former configuration (310, 350, 390) pass to become eastwards in the new architecture (see Fig. 8.2), complicating the transition period. RVSM introduction started in the North Atlantic Region in March 1997 and was developed in 13 phases until its end in November 2011, with the addition of Russian and Iraq FIR (Flight Information Region) airspaces.
- The transformation process was made by ICAO airspace regions, starting with the highest traffic and best equipped North Atlantic and finishing by Russia in 2011, leaving practically only the Antarctic airspace in the former status. Having more levels to choose, the airspace capacity increases and airlines can do a better optimization of the flight plans, according to meteorological conditions and other traffic circumstances. The benefits were quantified in savings of 310,000 yearly tonnes of fuel.
- *Free Flight*, when the flight is free and direct, without using airways or navigation from VOR to VOR (Very High-Frequency Omnidirectional Range) and not needing any indication from ground. This is the future scenario of the civil aviation world for the next couple of decades. The aircraft will be able to determine their position in flight with a high level of accuracy, using only satellite indications. Several constellations of satellites (the American GPS, the European Galileo, the Russian GLONASS, and the Chinese program BeiDou) will provide the information to on-board systems. At this moment, GPS and GLONASS are fully operative, with global coverture; BeiDou is only available in China and the Asia-Pacific region, but will reach worldwide range by 2020; Galileo is in initial test phase that will be finished around the same year than its Chinese counterpart. Collision avoidance will be exclusive responsibility of the aircraft Advanced Collision Avoidance System (ACAS) and the role of

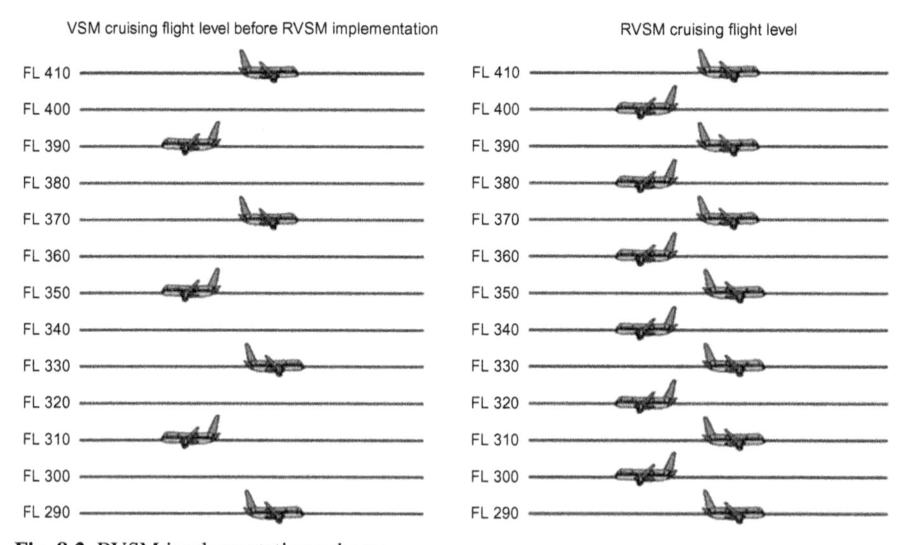

Fig. 8.2 RVSM implementation scheme.

ground stations will be reduced to monitor the correct performance of the global system. In this scenario, airways and waypoints will not be needed anymore and all flight will move along the shortest route. Potential fuel savings go up to 15% of present consumption, contingent on the amount of traffic in the different areas.

International routes can be improved with the collaboration of the overflown States. The most common procedures are to align the waypoints with the shortest trajectory or eliminate the some of the intermediate reference points. In Fig. 8.3, an example of the work made in the Frankfurt-Hong Kong route shows a benefit of reducing 31 min flight time per operation, equivalent to 3900 kg of less fuel consumption in a Boeing B777/200 aircraft.

The application of the best technologies to intercontinental flights requires the collaboration of multiple stakeholders of different States. The international pressure to reduce CO_2 emissions (and consequently fuel consumption) has fostered the launching of different initiatives in different world areas, like Atlantic Interoperability Initiative to Reduce Emissions (AIRE) in the North Atlantic in 2007. The equivalent Asia-South Pacific Interoperability Initiative to Reduce Emissions (ASPIRE) was formed in 2008 with participation of Australia, New Zealand, and USA at the beginning, joined later by Japan, Singapore, and Thailand. Three years later, Australia, India, and South Africa set up the Indian-South Pacific Initiative to Reduce Emissions (INSPIRE).

As a good example, AIRE was established through an EU-USA agreement, and developed initially by a consortium of 15 ANSPs, system manufacturers, engineering companies, and airlines during the first phase in 2008–09, trying to optimize oceanic operations. In the second phase in 2010–11, with the participation of 43 entities, the scope was widened to include all and each one of the different sectors of a commercial flight between the two continents (Fig. 8.4).

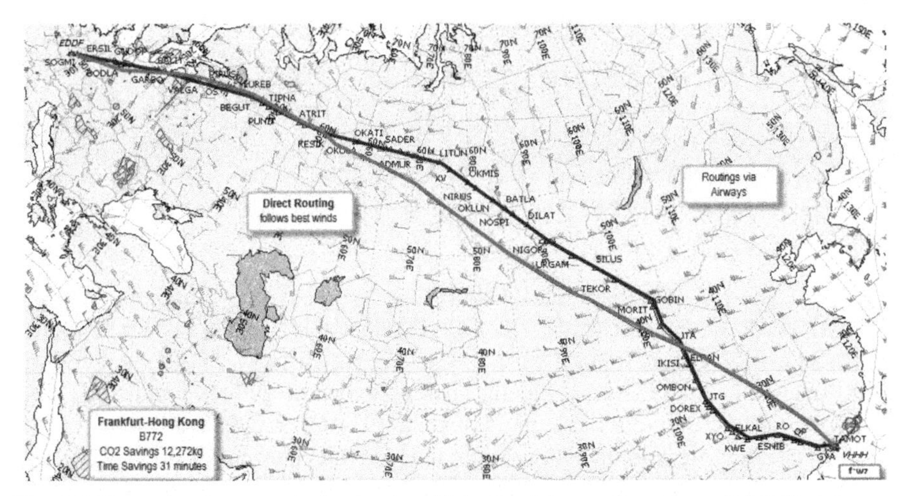

Fig. 8.3 Example of route optimization in a flight Frankfurt to Hong Kong (*IATA Training: Aviation and the Environment*).

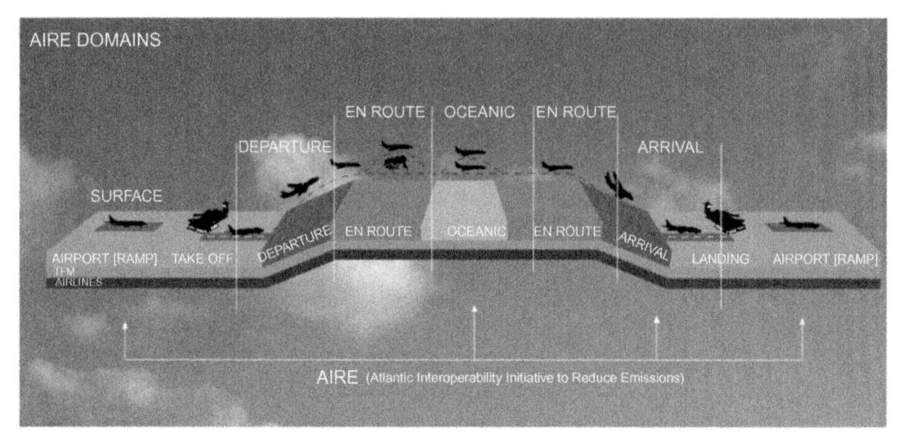

Fig. 8.4 Analysis of a route in the AIRE program (*SESAR: AIRE, Summary of results*).

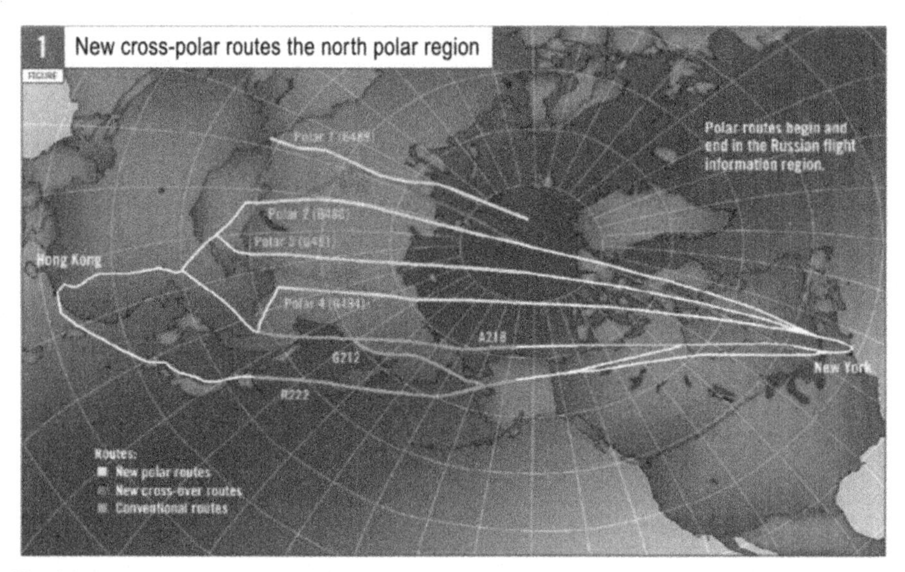

Fig. 8.5 Development of new Polar routes (*IATA Training: Aviation and the Environment*).

A precedent for all these multinational actions was the Russian-American Coordination Group for Air Traffic (RACGAT), created in 1998 to open new routes over Siberia or China and North Pole, shortening the distance for flights between North America and the Far East. Satellite-based navigation guidance and some specific measures to prevent fuel freezing at high cruise temperatures made them viable (Fig. 8.5) and very successful, cutting flight time and fuel consumption.

Other interesting approach to increase fuel efficiency improving navigation procedures is the integration of several national ANSPs in a single organization with common equipment and procedures, eliminating the differences among airspace management in neighbor countries. This is the EU approach with the Single European Sky (SES) program that intends to move from an airspace with 40 ANSPs to a single unified ATM region.

The Public Private Partnership (PPP) consortium SESAR (Single European Sky Advance Research) develops the technical part of the program with the following targets:

- Increase three times the European air space management capability
- Increase safety by a factor of 10
- Reduce 50% the ATM cost to the users
- Optimize flight trajectories to save between 8 and 14 min per flight, reducing fuel consumption by an average of 300–500 kg

The program started in 2006 and it is assumed to last until 2025. It includes the launching and put into service a global navigation satellite system, named Galileo, to provide a highly accurate, guaranteed global positioning service, interoperable with the similar US system (GPS) and Russian system (GLONASS). The satellite constellation will have 24 operative units and 6 spares and its entry into service is expected in 2020. Galileo was initially developed by the European Space Agency (ESA), a multinational State scientific body. In July 2017, ESA transferred the formal responsibility for oversight of the Galileo operations and provision of services to a newly created entity, Global Navigation Satellite System Agency (GSA).

The transition from a highly fragmented airspace (see Fig. 8.6) to a single airspace will be done in several steps that will join neighbor States airspace in commonly

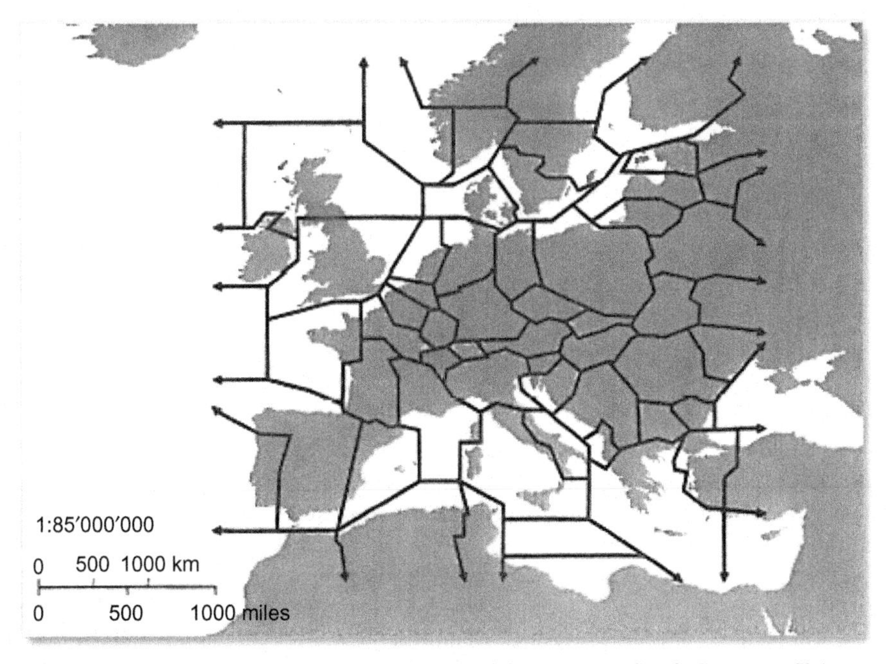

Fig. 8.6 ATM regions in the European air space (*EU Commission: Single European Sky*).

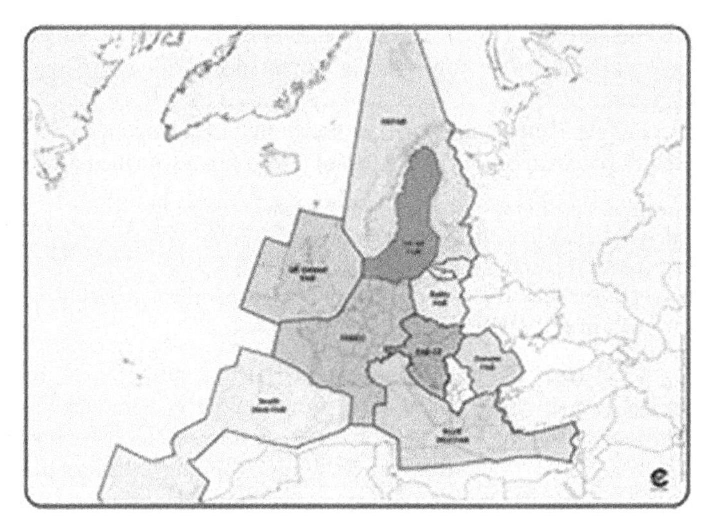

Fig. 8.7 The nine presently existing Functional Airspace Blocks (*EU Commission: Single European Sky*).

operated ATC areas. In 2014, a total of 9 Functional Airspace Blocks (FABs) were consolidated (Fig. 8.7), including 31 European States:

North Atlantic FAB:	Ireland, UK
Scandinavian FAB:	Denmark, Sweden
Baltic FAB:	Lithuania, Poland
Blue Med FAB:	Cyprus, Greece, Italy, Malta
Danube FAB:	Bulgaria, Rumania
Central Europe FAB:	Austria, Bosnia & Herzegovina, Croatia, Czech Republic, Hungary, Slovak Republic, Slovenia
FABEC FAB:	Belgium, France, Germany, Luxembourg, the Netherlands, Switzerland
North European FAB:	Estonia, Finland, Latvia, Norway
South West FAB:	Portugal, Spain

FAB definition has been made as a function of the existing airspace national limits and is not equivalent in terms of traffic volume. FABEC is the most important one, with a 55% of the whole European traffic crossing it.

An already-proven precedent, at small scale, is the Free Route Airspace Maastricht (FRAM), a program run by the Maastricht EUROCONTROL Centre, which gives air navigation services in the upper airspace (over 25,000 ft. altitude) of Belgium, Luxembourg, the Netherlands, and the Hamburg FIR. FRAM has opened 142 new direct routes since 2011 with 12,000 t of CO_2 savings per year.

United States has already a similar satellite constellation, GPS, that is operating successfully, but is also involved in an ambitious program of ATM improvement, titled NextGen, in order to speed up the transition from former Communication, Navigation,

Surveillance (CNS) to the modern satellite-based technology. The NextGen program started in 2003 and its implementation is increasing airspace capacity, improving safety, and reducing flying time, with accumulated results until the year 2018, of a 35% delay reduction and 14 million tons CO_2 savings.

The combination of SESAR and NextGen can offer fuel savings in the order of more than 1.5 million ton per year, according to IATA calculations, not later than 2021.

8.4 ATM efficiency evaluation

All of the aforementioned initiatives have the target of improving the efficiency of the ATM system in different issues like safety, fuel consumption, reliability, capacity, environmental impact, and interoperability. Projected improvements have to take into account the traffic evolution in global and local terms and the technology progress, in flight, and ground-based equipment. The evaluation of the ATM system efficiency has to consider the positive effects of technological and organizational efforts and the negative weight of traffic growth that asks for additional growth of capacity.

The question of how efficient is the actual ATM, compared with an ideal condition of everything working exactly as intended, is not easy to answer. At the beginning of the GANS study, ICAO considered that 2010 ATM efficiency was between 87% and 90%, meaning that, on average, every operating flight was consuming between 10% and 13% more fuel than it needs to. The traffic forecast associated to this analysis showed that the number of flights would double in the 2010–30 period. Imposing this growth in traffic on the 2010 ATM system, without any improvement, would result in an efficiency degradation of 0.2% per year or a total of 4% in the 20 years covered by the study.

ICAO figures do not coincide with the analysis made by the Civil Air Navigation Services Organization (CANSO), an association of the main world ANSPs. In 2012, CANSO published a document with an evaluation of tentative ATM Global Environment Efficiency Goals in 2050. The baseline year was 2005 and the global ATM efficiency was considered to be between 92% and 94%.

The main difference of both evaluations was the approach adopted. While ICAO accounts for all the different factors in a package, CANSO makes a distinction between effects that have direct repercussions on fuel consumption, like the flown distance, and other indirect elements, such as insufficient airport capacity, grouped under the name of interdependencies. A schematic representation of this concept is in Fig. 8.8.

The interdependencies identified in the CANSO report can be divided into seven categories:

- *Airline practices*, because not all the operators have flight planning systems with the capability and flexibility needed for taking advantage of all the most optimum routings that may be available. These deficiencies may be organizational or coming from the technological level of the operating aircraft and support equipment.
- *Capacity*, when capacity limitations appear, an aircraft may be required to hold in-flight, waiting for an available slot or wait during the taxi out before take-off. ATM system has the possibility to increase the airspace capacity but has no authority on airport capacity or on the slot allocation procedures.

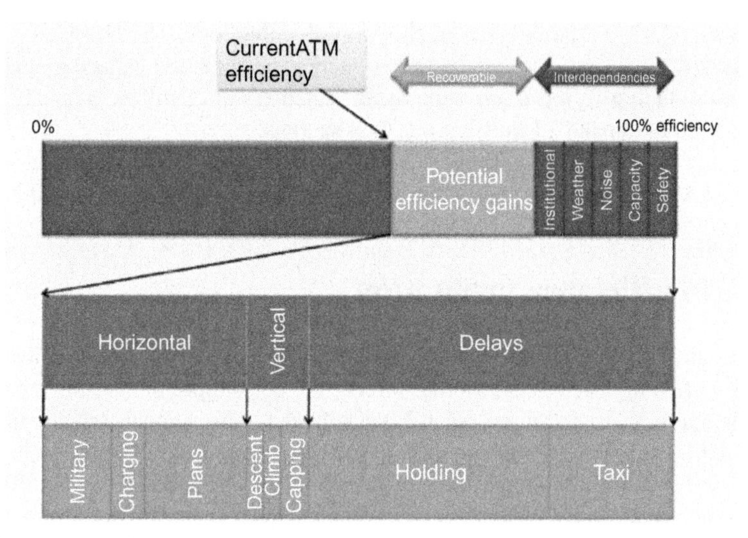

Fig. 8.8 ATM potential efficiency gains and interdependencies (*CANSO: ATM global environment efficiency goals for 2050*).

- *Institutional*, if fragmented airspace constrains flight planning. Different regions or countries may have different noncompatible operating procedures, requiring deviations from the optimum conditions. Mention should be made on charging systems: if air navigation charges are very different, the operator might be tempted to make longer flights for using the cheapest airspace. For example, under EUROCONTROL charging system, the 2017 unit prices in different airspaces vary from the most expensive area (Switzerland) to the cheapest (Portuguese Azores Islands FIR) in a ratio of 10 to 1.
- *Military*, as it was previously discussed, military-restricted air zones, permanently or temporarily, require close cooperation between civil and military ATM, in order to optimize the trajectories.
- *Noise*, if noise abatement procedures, specific for the operation in some airports, are different than the minimum consumption ones. It depends on the location of noise sensible areas around the airport and used to be more problematic for take-offs, if the tracks try to avoid some populated areas. Most modern aircraft have FMS equipment with automated climb procedures for either minimum fuel or minimum noise trajectories. In the case of approach and landing, the CDA procedure is able, in most of the occasions, to reduce both noise on ground and fuel consumption at the same time.
- *Safety*, as a flight can be deviated from the optimum track in order to ensure adequate separation from other close aircraft. This separation can be dictated by aircraft types (wake vortex) or by the accuracy level of the local ATM facilities.
- *Weather*, if avoidance of adverse weather systems may recommend nonoptimum routing to ensure a safe and smooth flight.

There are a number of good examples in which interdependency effects can be drastically reduced by effective collaboration among different stakeholders. The RVSM operation, described in the previous chapter, is a success story to take into account.

A different and more aggregated evaluation is in the 1999 IPCC report, giving a range of 6%–12% fuel reduction per trip thanks to potential ATM improvements, implemented in the 2000–20 period. The achievement of those figures is considered contingent on the implementation of a number of essential institutional arrangements at an international level. The analysis does not consider any airport or other infrastructure capacity constraints.

In 2014, ICAO performed a study of the potential efficiency results of the ASBU Block 0, the finalization of which was scheduled for the end of 2018. The most likely results point to something between 2.3 and 4.1 million tonnes of fuel savings per year, considering the level of predicted implementation of the different proposed measures. With an oil price around 60 USD per Brent barrel, that will provide 1200–2200 MUSD yearly savings to the operators.

The results indicated that a fuel burn reduction of 2%–3% was possible if the totality of the measures listed in Table 8.1 was implemented worldwide in that time period. Fig. 8.9 illustrates the difference of both cases, always considering the 1% of efficiency degradation by the traffic increase in the 5 years, 2014–18. The immediate conclusion was the insufficiency of the present pace of improvements to improve substantially fuel efficiency but, at the same time, the proposed measures show good potential if implemented faster and over a wider number of regional airspace zones.

The evaluation of efficiency improvements in local air spaces is a complicated task, prone to offer many different results when nonhomogeneous methodologies are used. However, individual states have to offer some indicative results of the different initiatives implemented in their sovereign airspace. As an example, Table 8.2 shows a year 2013 quantification of the major initiatives taken by India in its airspace.

In order to make those analysis more comparable each other, ICAO has developed a Fuel savings Estimation Tool (IFSET) to assist States to estimate fuel savings in a manner consistent with CAEP approved models and aligned with GANP.

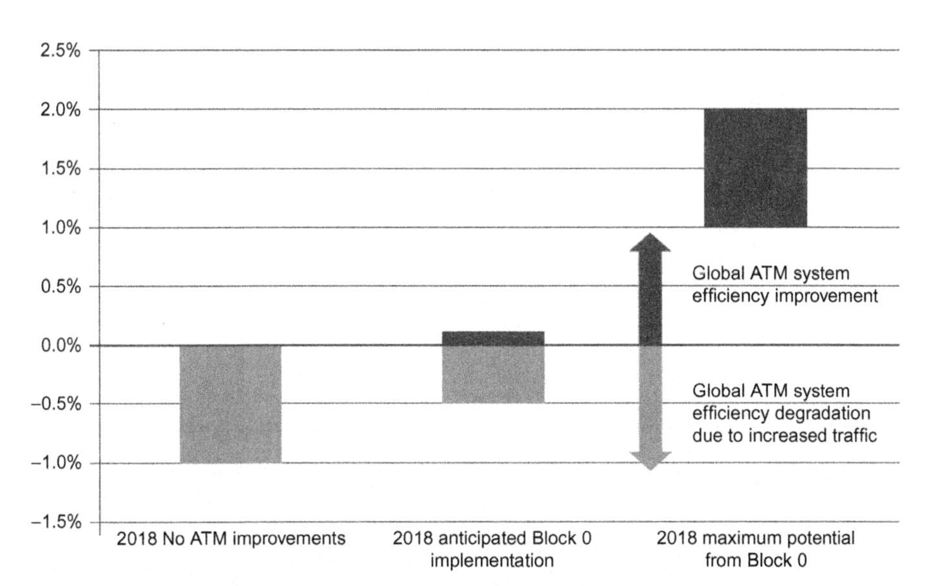

Fig. 8.9 Expected Fuel savings goal after Block 0 implementation (*ICAO: Air Navigation Report*).

Table 8.2 Consolidated estimated benefits from the major initiatives taken in India (*ICAO: Air Navigation Report*)

ANS improvements	Fuel saving (per year in tonnes)	Cost savings (per year in million $)	Remarks
50 NM RHS	104,573	114.98	16 routes
RNAV 5	14,637	16.06	Q1 to Q13
New domestic routes	9889	10.95	8 routes
RNP 10	11,662	12.78	L875,756,516,899,518
Three runway ops	13,140	1.30	Delhi
Upper Space Harmonization	18,060	19.90	Chennai FIR
INSPIRE	218	0.20	Based on 1031 UPR flights
PBN	22,863	25.11	Based on 6 Airports
Enhanced Surveillance	14,500	16.00	RHS on W20 and R460
CDO	1164	1.30	Based on Ahmadabad ops
Connector routes	4095	3.65	V1–V32
Total	213,610	222.23	

IFSET methodology is to evaluate the differences in flight trajectory performance in terms of fuel consumption before and after implementation of operational improvements in the analyzed area. The calculation covers four different categories of improvements:

- Reduced cruise distance or time
- Availability of optimal altitude, requested for the flight plan
- Reduced taxi time
- More efficient departure and approach/arrival procedures

This methodology has a highly simplified aircraft characterization, in terms of weight, thrust, CG position, and airframe/engine combination. Then, it is not appropriate to calculate actual fuel consumption and its best use is to give a comparison between any two scenarios. A number of analysis have been made for individual countries (like India) and for associations of State ATMs, like ASECNA and COCESNA.

8.5 Airport energy management

The contribution of airports to aviation energy efficiency has three main aspects, representing the different kind of airport activities. A first area of action is the relationship with the aircraft operation in the airside of the facilities, regulating aircraft operation inside the Airport Terminal Area, a volume of about 3000 ft. height over the ground. In addition, runways, taxiways, and local circulation zones create a space where aircraft interact with a great variety of ground vehicles in the maintenance and turnaround maneuvers.

A second element is the groundside airport installations, like terminals, service areas and different buildings with activities related with the airport, as hotels, shops, warehouses, car rentals, and a long list of resources serving passengers and airport personnel. Their size used to be relatively large as the amount of passenger and cargo grows. Some of these facilities may not serve directly the airport but take advantage of the activity generated by it for providing supplementary goods and services.

The third group comprises the airport surroundings and the transportation modal exchange facilities. The different ways of accessing the airport by passengers and employees are often totally or partially controlled by several entities different from the airport management, and may have additional uses, other than serving the airport itself. Ground infrastructure for roads, railways, or underground trains has importance on the energy consumption and local air quality, as well as the regulation of the vehicles entrance, allowing more or less private cars, taxi, buses, short- or long-range trains, etc.

It is relevant to note that the Airport Authority may have more than one purpose for its energy policy. Minimizing energy consumption is one of them but, at the same time, the variety of available energy sources (electricity from different origins, fuel, gas, geothermic) allows a certain economic optimization of the supply mix. In this selective process, the environmental factor has also weight, because reducing the carbon footprint is becoming more and more important.

Energy mix depends on the area where the airport is located, and the management policy of consumption, many times related to environmental footprint concerns. The majority of final consumption used to be electricity, but there may be differences in the sources of that electricity. In Heathrow airport, for instance, 87% of its energy budget is electricity provided by the general network, followed by 9% of gas, 3% of biomass, and 1% of gas oil. Other facilities may use their particular situation to obtain electricity directly. In 2015, CIAL (Cochin International Airport Limited) in Kerala, India, became the first in the world to be fully powered by solar energy, thanks to a 12-MW solar farm, placed on an 18-ha site, close to the cargo terminal, to be added to other installations on the roof of the arrival terminal and the maintenance facility.

In a different development, La Palma airport (Canary Islands, Spain) obtains its whole energy from a 1.3-MW wind energy plant (Fig. 8.10), since July 2003. Some minor fuel consumption emitters like emergency energy generator or access vehicles are offset by excess wind generated electricity distributed to the city network. This type of installation needs a careful spatial design to avoid interferences and light reflections from the blades into the aircraft operations.

Some airports can negotiate with their energy providers and specify the source of their electricity. The Schiphol group in Holland will use only wind-generated electricity in its four airports (Amsterdam, Eindhoven, Rotterdam, and Lelystad), starting January 1, 2018, after closing an agreement with the energy company Eneco. Annual consumption is about 200 GWh. In other cases, energy options are dictated by strategic reasons but may have efficiency repercussions as well. Cook Islands have a plan to rely totally on renewable energy by 2020 for its electricity coverage and reduce the country's dependence on fossil-imported fuel. A part of this plan is the installation of 3051 photovoltaic panels in Rarotonga International airport, producing up to 960 kW, about 5% of Rarotonga's energy needs, and avoiding the import of 400,000 L of Diesel fuel per year.

Fig. 8.10 La Palma airport obtains its energy from wind generators (Hesse, José Manuel: Climate Change impact of airports and air navigation activities).

Airport airside

The energy savings in the flight operations within the airport Terminal Area, like the use of CDA/CCO procedures, have been treated in Chapter 6. Therefore, this part will cover only the ground operations, divided into four parts:

- *Managing the taxi time of the aircraft*: The airport configuration has a great influence on the distances that aircraft has to cover from the runway to and from parking positions. Although the geometric configuration is defined at the project of the facilities, daily allocation of gates and parking slots may help to reduce the fuel burn on the ground.

  Originally, commercial aircraft taxiing procedures were designed using all engines at a low thrust, typically around 7%–8% of the maximum certified thrust, used for high-weight take-offs. A step forward was turning off one engine (or two in four-engine models) for the taxi in until reaching parking position, saving a nonnegligible amount of fuel in that way. The symmetric maneuver for the taxi out until take-off position is not generally used because pilots are afraid of a possible failure at the start of the inactive engine that would force the aircraft to get back to the parking position.

  There are a number of technical developments with the purpose of doing the whole taxi maneuver with all engines out. For example, the Electric Green Taxi System (EGTS) International is a joint venture of Safran and Honeywell Aerospace proposing the installation of a small electric engine in the nose landing gear of the aircraft (Fig. 8.11).

  The engine will be fed by the APU and controlled by the crew, allowing not using engines during taxi. The negative part is the additional weight of the electric engine, but in short/medium-range flights of an A320 size aircraft, it may reduce fuel consumption up to 4% and about half of the local air quality emissions of the taxi maneuver. In addition, it offers the advantage of an autonomous pushback, saving handling costs.

  A more classical approach to the problem is the TaxiBot, an hybrid-electric towing vehicle designed by IAI (Israel Aerospace Industries) and Lufthansa Technik AG. It is an 800-hp

Fig. 8.11 Electric engine for autonomous taxi in the A320 nose landing gear (Hesse, José Manuel: Climate Change impact of airports and air navigation activities).

vehicle, controlled by the crew (Fig. 8.12). The prototype is being tested in standard service in Frankfurt airport.

- *Energy supply to the parked aircraft*: The majority of commercial aircraft are autonomous in the use of energy when parked, either keeping engines running or using their APU to generate electricity and air conditioning. APU consumes a high amount of fuel and it can be replaced by other energy sources more efficient in the production of the energy type needed. Those may be Ground Support Equipment (GSE) or direct connections to a centralized electricity network or an acclimatization central.

Fig. 8.12 TaxiBot in Frankfurt Airport (www.lufthansa-leos.com).

- *Low-energy consumption GSE*: Aircraft service at the ramp conveys a number of vehicles around. Fig. 8.13 illustrates the turnaround of a Boeing B787 with up to 17 units around, including a towing tractor, a potable water service truck, an air-conditioning truck, an hydrant fuel truck, three ground power electrical generator units, a belt loader, a cabin-cleaning truck, a waste-servicing truck, two galley service trucks, two cargo loaders, and three baggage transportation trains. All these vehicles are designed to move around the airport ramp, doing short trips at moderate speeds, being ideal for using gas or electric engines and reduce fuel consumption. Liquid Hydrogen is other alternative for the heaviest vehicles.

  The same is applicable to people transporting units like passenger buses or staff cars. Many airports are introducing electric vehicles (Fig. 8.14) and implementing the corresponding modifications in the ramp, like installing charging poles with a monitoring system and establishing the standards for charging and payment. Electric vehicles are initially more expensive and the airport may incentivize their use through clauses in the concession contracts with the airlines or ramp operators.
- *Low-consumption lights in runways and in the ramp*: the replacing of conventional lamps by LEDs offers substantial energy savings and much better reliability and useful life. Heathrow airport finished this operation at the end of 2016 and estimated annual energy savings equivalent to 385,000 sterling pounds.

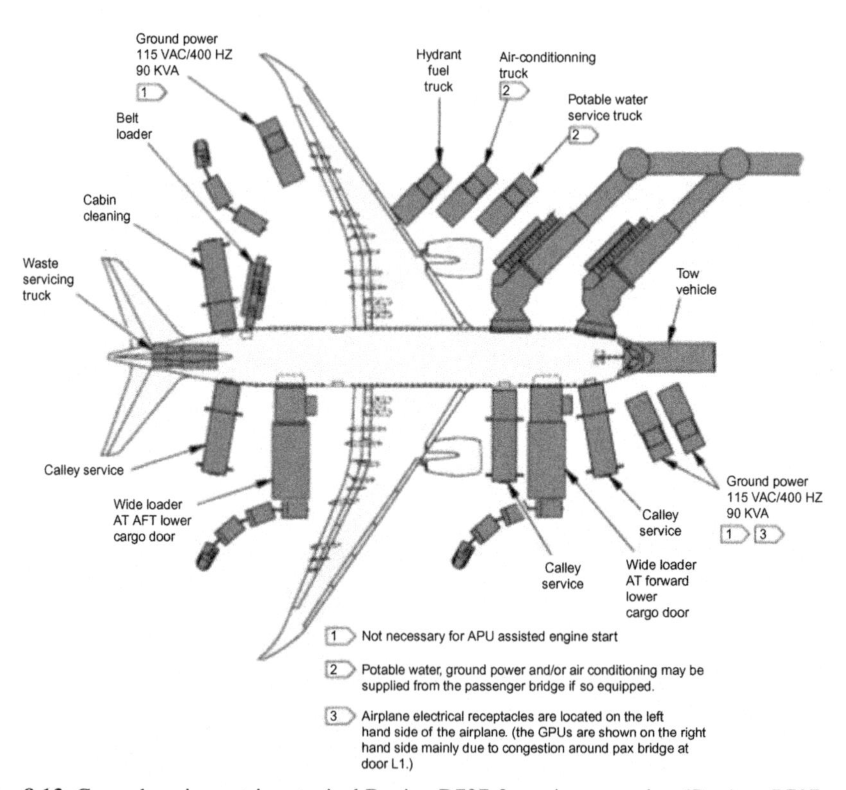

Fig. 8.13 Ground equipment in a typical Boeing B787-8 service operation (*Boeing: B787 Airplane Characteristics for Airport planning*).

Fig. 8.14 Electric ramp vehicles at Barajas airport (Hesse, José Manuel: Climate Change impact of airports and air navigation activities).

Airport ground side

The highest energy consumption in an airport is coming from the passenger and cargo terminals. The daily electricity and thermal energy used by a large airport can be compared to that of a city of 100,000 inhabitants. The distribution of that consumption changes depending on how developed is the nonaeronautical activity sector in each specific case. For a middle-size passenger terminal, about 70% corresponds to Heating Ventilation and Air Conditioning (HVAC), 15% to lighting, 12% to shops and concessions, and 3% to other different elements. Then, the climate control system of the building is very relevant for the global energetic efficiency.

The design of the terminal is determinant in the final result and it should be adapted to the type of environment in which the airport is located. Hot climate places ask for heat shielding and wise graduation of the incoming light. Singapore's Changi airport Terminal 3 has a roof with thousands of small rotating plates, controlled by an automatic system, in such a way that it allows to enter the amount of light needed in each time of the day, depending on the prevalent atmospheric conditions.

A typical efficient lighting system must comply with the following guidelines:

- Give priority to natural light
- Use of transparent elements to avoid opaque walls
- Photocells to control the turn on of the artificial lighting
- Selective lighting in area with little use
- Indirect lighting in transit areas
- Reduced intensity lighting in resting/relax areas

The big roof surface of the terminal buildings makes them excellent candidates for solar panel installation to generate part or even the totality of the electricity demands. Geneva airport has a step-by-step movement in that direction, started in 2006 with the installation of a solar power plant on the main hangar roof, followed in 2010 by two solar panel arrays on airport buildings, expanded in 2012 up to a total of 10,000 m^2.

The policy of Geneva airport is to sign contracts with the installation builders to buy the electricity instead of operating them itself. In 2017, the airport entered into a contract with public energy utility SIG (*Services Industriels de Genève,* Geneva Industrial Services) to install solar panels on 50,000 m^2 of roof space. The array will produce around 7.5 GWh of electricity per year and will be installed on the roofs of 10 buildings around the airport. SIG will build and own the panels for 25 years and sell the electricity to the airport. Other airports, like Denver or Charles de Gaulle, with less surface of flat roofs are creating photovoltaic plants, the so-called solar farms, in areas close to the terminals.

While the energy efficiency started with a thermally efficient design, the improvements along the following years are facilitated by the implementation of an Energy Management System. A good example is the Hong Kong International Airport terminals design. They use external modern glass and building envelopes to reflect heat and reduce cooling loads. Natural light income is optimized by roof skylights and light sensors automatically reduce indoor lighting when there is sufficient daylight. This type of gradual intensity control is very useful in a big terminal where service demands change along the hours of the day and requires some type of energy adjustment. For example, escalators can be equipped by motor controllers, which automatically reduce power to the motors when there are fewer people on the escalator, saving energy.

Geothermal energy is other possibility being used by a growing number of airports, in the majority of the cases for heating terminals and other buildings. In 2011, Paris-Orly airport put into service a geothermal system that provides 70% of the heating in the Terminals and some new buildings constructed in the Coeur d'Orly district. Two wells were drilled to a depth of 1800 meters to meet the heat requirements. The water comes up at 74 degrees Celsius through an overall district heating system 35 km long and travels through 108 substations. Airport gas consumption is reduced yearly by 4000 t of oil equivalent and avoid the emission of approximately 9000 t of CO_2.

A new technology that is used more and more frequently is the geothermal heat pumps, having the capability of raising or lowering the temperature of a working fluid and can therefore be used for both cooling and heating. The latest new terminal in Zurich airport, a building 30 m wide, is founded on 440 foundation piles, 30 m depth, of which 315 are equipped with heat exchanger tubes. The system extracts annually 1.1 GWh of heat for heating the terminal in the winter. In the summer, the building heat is deposited in the ground, which thus act as a storage medium. For air-conditioning, some 500-kWh cooling energy is supplied by these geo-structures or "energy piles."

An additional application in study is the possible use of geothermal energy to melt ice and snow on runways and airport ramp. At the moment, a combination of snowplow and chemical deicers is the standard solution. Snowplow machines are dangerous in the airport airside because they may not be seen by operating aircraft and, in many cases, and it is mandatory to close the area in which they are working. Chemical deicers may have undesirable side effects. Those based in urea (the most widely used) are toxic to many life beings while other inorganic chemical products may cause erosion in the runway concrete. Geothermal energy is a good candidate to replace them, especially in countries with a long experience in geothermal snow melting, like Iceland or Sweden.

Other alternative to the use of geothermal energy is the concrete plates with an internal electric conductivity system. Concrete has about 1% of its weight in carbon fiber that, when connected to an electric supply, gets hot. A 4.5×4.0 m plate consumes 333 W for deicing. The system is presently being tested in Des Moines airport and, in addition to the positive environmental impact due to the elimination of chemical products, has a positive energy benefit by taking out the snowplow machines.

Thermal isolation of the terminals is important in order to avoid convective energy losses, detecting the main heat transmission points between the building and the outside. Typical measures on this issue are:

- Insulating panels
- Replacement of thermally conductive materials
- Front part covering
- Paintings with insulating properties and adequate colors to avoid summer heating and keeping warm in winter
- Install shadow producer elements in the front part
- Use of local vegetation as climate control mechanism

A good example of these practices is The Queen's Terminal at Heathrow airport (see Fig. 8.15), recently built to replace the old Terminal 2. The target was a 40% reduction of the CO_2 emissions compared with the previous building. Approximately half of it was achieved by reducing the energy consumption and the other half by the use of renewable energy sources.

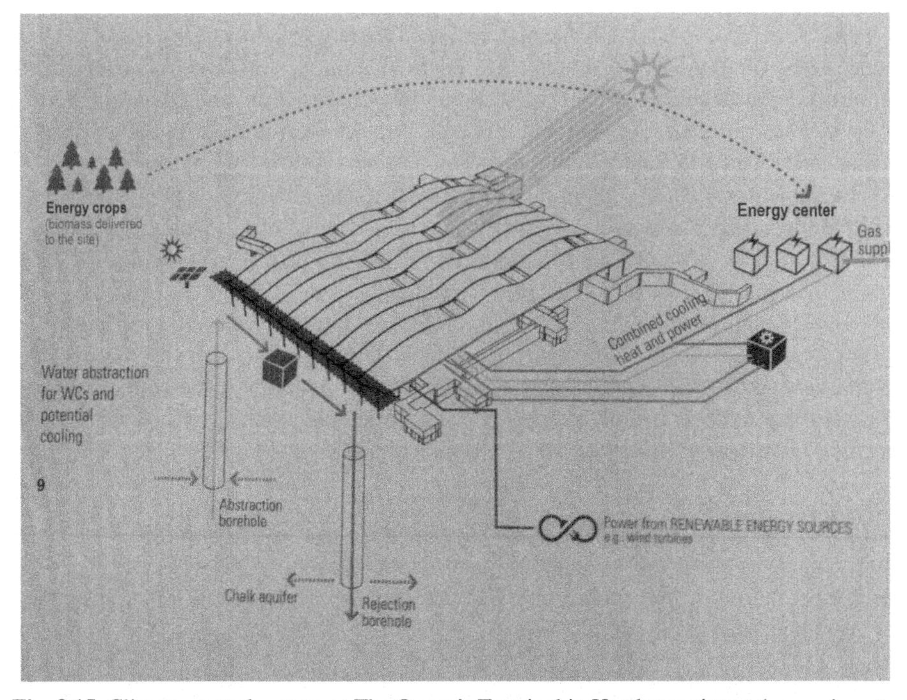

Fig. 8.15 Climate control system at The Queen's Terminal in Heathrow airport (*www.airport-technology.com*).

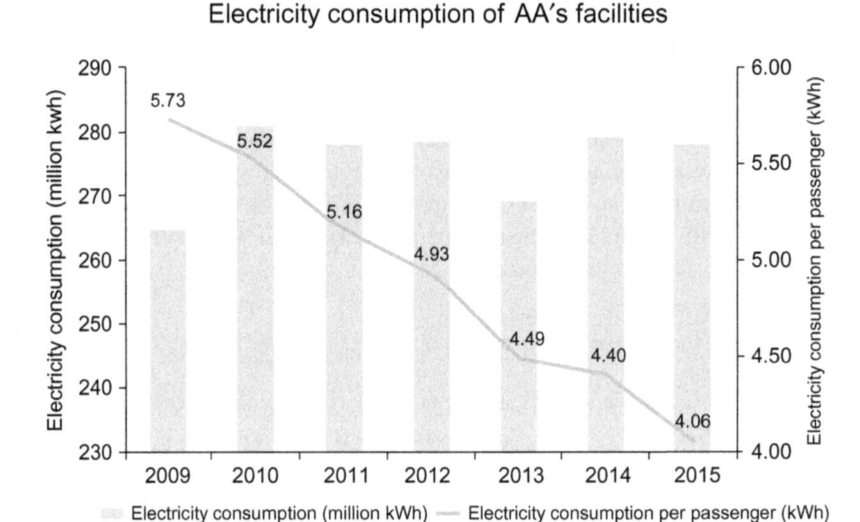

Fig. 8.16 Electricity consumption evolution in Hong Kong airport (https://www.hongkongairport.com).

The ample space of those huge buildings makes difficult a reasonable temperature management without using a lot of energy. In Hong Kong airport, innovative cooling systems are in place to cool only the bottom 3 m of the large indoor spaces, while leaving the air above at ambient temperature. Key environmental performance indicators were identified and carried and carried out in the design, construction, and operation phases of the new infrastructure. In February 2017, the airport received the ISO 50001 Energy Management Certification for Terminal 1. This standard requires organizations to establish, implement, maintain, and improve their energy management systems to identify energy reduction opportunities that lead to continuous improvement in energy efficiency. A sample of this continuous improvement can be seen in Fig. 8.16. During the 2010–15 period, the electricity consumption has been stable, between 270 and 280 million kWh, although the number of passenger passing through the airport increased by 35%.

With respect to the shops and other concessions, the energy they use is provided by the central services, but the airport Authority may put some pressure on the business operators, inserting in the rent contracts some clauses demanding an energy-efficient behavior and establishing some efficiency targets related with the level of rent they are paying.

Airport access

Public transport is the most energy-efficient transport mode to go to the airports, but not always the most convenient for the traveler needs. Classified in terms of energy savings, there are four main categories:

- *Subway*: the best energy consumption per passenger and having a wide network of connections and accessibility places. Generally is low priced. On the negative side, it takes a lot of time due to many intermediate stops and is not friendly for passengers with large and heavy baggage. In many cities, it is not a 24-h service and does not provide service for late night or early morning flights.

- *Train*: dedicated trains are the fastest and more convenient access way to the airport but usually are more expensive than subway or bus. Some of them are almost exclusive for airport traffic, like Heathrow (almost 25% of all passengers use it) or Narita Express, and have few downtown stops to board them. In many European airports, the trains stop at the airport as one of several points in their schedule, mixing airport-city trains with general railway services, like in Brussels or Geneva. In some cases, even high-speed trains (HST) have a stop at the airport and allow passengers to perform intermodal connections with more distant places, like the French HST *Thalys,* transporting passengers from Charles de Gaulle airport in Paris to the *Gare du Midi* railway station in Brussels.
- *Bus*: the cheapest and, generally, slowest of all the access means. Like the subway, it has a wide coverage area but many intermediate stops may make the trip very long. It is better than subway for travelers with baggage, but in both cases, a seat may not be guaranteed.
- *Taxi*: the most expensive and the least energetically efficient. Easy to access with or without luggage. Depending on the road traffic situation may be slow sometimes. In many cities, there is a flat fare between downtown and the airport to help visitors to be sure about the trip cost and not suffer abuse by some drivers.

The airport management attributions to regulate ground traffic access to the facilities uses to be limited and in the majority of the cases, airport, transport, and municipal authorities must negotiate and achieve an agreement acceptable for everybody. The results are very different according to the general transport policy of each State or local community. Table 8.3 shows a comparison between the big airports in USA and those in Europe, with some Chinese and Japanese facilities. The percentage of passengers

Table 8.3 Use of public transportation by travelers going to the airport (*Coogan, Matthew A.: Ground access to major Airports by public transportation*)

US airport	% of Public transport	European airport	% of public transport	Asian airport	% of public transport
San Francisco	23	Oslo	64	Hong Kong	63
New York-JFK	19	Zurich	47	Tokyo-Narita	59
Boston	18	Vienna	41	Shanghai	51
Washington-Reagan	17	London-Stansted	40		
Oakland	15	Paris-C. de Gaulle	40		
New Orleans	15	Amsterdam	37		
New York-Newark	14	Copenhagen	37		
Atlanta	14	Munich	36		
Denver	14	London-Heathrow	36		
Los Angeles	13	Stockholm-Arlanda	34		

using public transportation is much higher out of the USA, as cities provide a better ground system to reach the airports.

In addition to the passengers, a great number of the transportation service users are not travelers, but employees working in the airport in aeronautical or nonaeronautical entities. This demand is more regular, with peaks at the times of starting and finishing the working periods. There are some ways to encourage the use of collective transport, like including bonus package to public transportation companies in the worker salary, promote car-pooling, or even offer bus services exclusively for airport workers.

The airport monitoring and controlling tools are a number of indirect policies and incentives to encourage the use of public transport that, in addition to the energy savings and reduced local air quality contamination, helps to keep low the access roads congestion.

An example is the Swiss airport of Geneva that has a well-defined plan to reach 45% of passengers and staff using clean public transport means by 2020. The plan has some key elements:

- Arriving passengers are provided with free tickets to go downtown either by train or bus.
- Trains go directly from arrivals terminal to downtown Geneva and some of them continue trip to the main Swiss cities and some other international destinations.
- Municipal bus stops are very close to the terminal at standard fare.
- Only low-emission taxi vehicles (electric, gas, hybrid) are allowed to pick up passengers in the airport.
- Staff car-pooling is encouraged and facilitated by a dedicated website

The restrictions to less efficient and more pollutant taxi models are increasing in recent years. Stockholm Arlanda airport has introduced a computer system to give preference to the taxi models with the highest fuel efficiency and the lowest environmental impact. To this purpose, a biogas filling station, the largest in Sweden, has been open at the airport. The number of "ecotaxis" arriving at the airport has increased from 47% to 87%, reducing carbon emissions by over 9000 t a year. Geneva airport is being more drastic and only allows electric or hybrid taxis to get into the airport.

Other way of adjusting offer to demand and save fuel is the use of smaller, energy-efficient vehicles. The company Cab4one offers a low-price trip from central London to Stansted airport to single travelers and their baggage in a two-seat Smart car.

Parking availability is other of the most powerful tools to regulate car access, but not all the airports move in the same direction. Some of them, like Zurich, apply very high fares to discourage the use of private vehicles or, like Terminal 5 in Heathrow or British Airways Headquarter, close to that airport, restrict the number of parking places available to the employees. However, other airports, like Adolfo Suarez- Barajas, in Madrid, offer low parking prices as an additional incentive to the travelers.

8.6 Airport environmental and energy conservation certification

The particular structure of the airports, with their combination of aircraft operations, immigration and customs systems, public service buildings, and commercial centers makes difficult the application of very specialized environmental and energy conservation standards.

A number of airports apply general scope International Standards Organization (ISO) recommendations, like ISO 14001 for Environmental Management Systems or ISO 50001 for Energy management Systems. There are also regional standards like the European EU MAS Eco-Management Audit System or those specialized in a part of the airport, like the US & Canada Leadership in Energy and Environmental Design (LEED), applicable to the buildings.

In 2009, Airport Council International Europe (ACI-Europe) launched an initiative, called Airport Carbon Accreditation Program (ACAP), to help airports to reduce their carbon footprint end eventually move it until zero value. The program was initially supported by the European Union, EUROCONTROL, and the European Civil Aviation Conference (ECAC) and was expanding progressively to other continents. At the end of October 2017, 201 airports of 61 States were participating in the program, moving about 40% of the world passenger traffic.

Although the purpose of the program is focused on emission reduction, with a special attention on greenhouse gas emissions, like CO_2, the structure of the ACAP has many coincidences with energetic efficiency programs and, as a consequence, airports are moving the two issues in parallel, applying the same procedures.

ACAP is based on a typical continuous improvement management approach, creating four levels of excellence to be reached in successive steps (in brackets the number of airports having achieved that level in October 2017):

- *Mapping* (58): Determine emission sources, calculate annual emissions, and compile a carbon footprint report to be verified by an officially recognized independent expert. This part, already fulfilled by the 201 airports participating in the program, requires to identify the energy consumption of the airport-controlled functions, by type of emissions and determine the best action procedures to reduce carbon footprint, including energy load reductions and/or changing the source of energy supply to more environment-friendly substances. In general, airports include here some reduction targets that present no great technological or organizational problems, and implement basic management tools to be developed later on, trying to show early and not very costly program successes.
- *Reduction* (65): Provide evidence of effective carbon management procedures and achieve specific reduction targets. Up to 143 airports have reached this stage in which management tools have been consolidated and the airport has achieved its targets in the processes not dependent on third-party activities.
- *Optimization* (43): Include third-party emissions at the airport (airlines, handlers, etc.) and around airport (surface access providers). In the case of aircraft operations, only LTO emissions (below 3000 ft. height) are accounted. Only 78 airport companies have managed to get agreements with the other parties operating at the airport area, who have accepted to be involved in the ACAP system, with the same level of accuracy than the airport itself. The great majority of the compromises are achieved through specific clauses in the contracts, asking for certain energetic efficiency standards, related to the prices or even the concession of the service.
- *Neutrality* (35): Neutralize remaining direct carbon emissions by offsetting through internationally recognized offsets (such as EU allowances). The last part of ACAP has nothing to do with efficiency but with the way to compensate emissions impossible to eliminate due to technical reasons or too costly for the airport operator. In this case, the neutrality is achieved by buying emission permits, the price of which is used to eliminate carbon emissions in other activities at much lower cost.

Energy efficiency and the environment

9

9.1 Environmental impact of air transport

Despite all the benefits of air transport to our societies, its environmental impact is becoming a concern from different perspectives. The fuel utilization has different environmental impacts: local air quality, consumption of nonrenewable materials, and contribution to climate change.

From a broader perspective, the environmental impacts of air transport are traditionally classified according to their reach in local and global effects. Local effects are restricted to the airport area, while global effects have a worldwide scope. The main impacts are:

- Local effects
  - Noise
  - Local air quality
  - Land and space use
- Global effects
  - Nonrenewable materials consumption
  - Climate change contribution

Concerning local effects, aircraft noise is by far the impact causing more complaints from affected communities. For instance, European airports receive one local air quality complaint per each 300 noise complaints. Engine noise is dominant under the takeoff and initial climb path and along the sides of the runway as well. Climb gradient is the key element for the noise impact. In approach and landing, engine and airframe noise are comparable. They depend on the aircraft configuration.

ICAO requires an acoustic certification before granting the Type Certification to new aircraft types. The applicable regulation is in Annex 16, Part 1, to the Chicago Convention. The regulation stringency is increased as the acoustic technology advances in order to ensure the application of the best-available technology.

The use of land and space is related to the following elements:

- Land and infrastructures (airports and air navigation control centers).
- Air space for flight airways, distributed according to the local air navigation service provider capabilities.
- A frequency band of the radioelectric spectrum for communications and ATC services.

This effect is relatively modest compared to other transport modes. Airports (including the airport ramp, runways, and the aeronautical and environmental domains) take about 1% of the overall land used for transport infrastructure. Competition for air space is comparatively lower. There are protected areas due to military use, security, or

Energy Efficiency in Air Transportation. https://doi.org/10.1016/B978-0-12-812581-6.00009-0

wildlife care. The present Air Navigation System is based in airways, but may evolve toward free flight with satellite guidance as new satellite constellations (Europe, China) join existing GPS and GLONASS.

Civil aviation uses a number of exclusive frequencies in the radioelectric space for voice communications and data transmission (according to ICAO-ICU agreements). In terms of allocation, civil aviation communications are divided into two groups: those needing high integrity and fast response (by safety reasons) and those corresponding to administrative issues or passenger service. Some frequency bands (VHF COM) might become saturated in the near future.

Local air quality in the airport area is regulated by the general air quality law of each country or region and it should include the impact of every activity within the airport area (Fig. 9.1). The European Union has a general rule establishing pollutants limits, but each Member State or local Administration may apply more stringent values.

The two major products of fuel combustion are carbon dioxide, CO_2 and water, H_2O. Other products of fuel combustion are nitrogen oxides, NOx; sulfur dioxide, SO_2; carbon monoxide, CO; unburnt hydrocarbons, UHC; and Soot (Fig. 9.2). Although the amount of species produced in the combustion of 1-ton kerosene depends on parameters such as the aircraft-operating conditions, altitude, humidity, and temperature, the following figures can be taken as good approximations:

- CO_2 3.15 ton
- H_2O 1.239 ton
- NOx 6–20 kg
- SO2 1 kg
- CO 0.7–2.5 ton
- UHC 0.1–0.7 kg
- Soot 0.02 kg

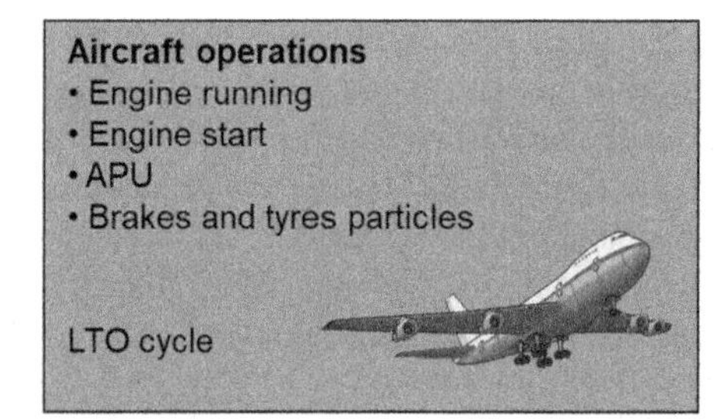

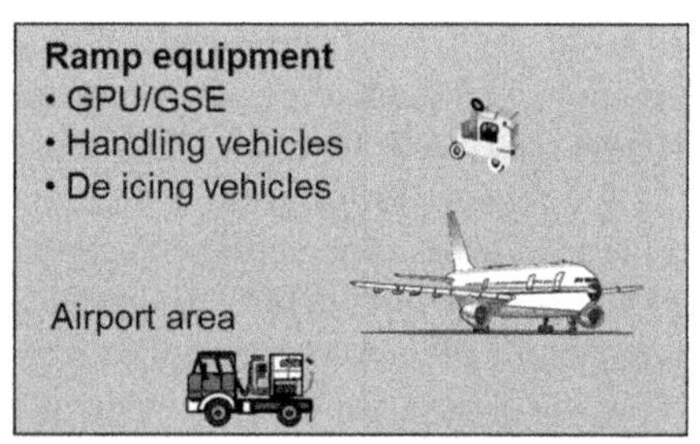

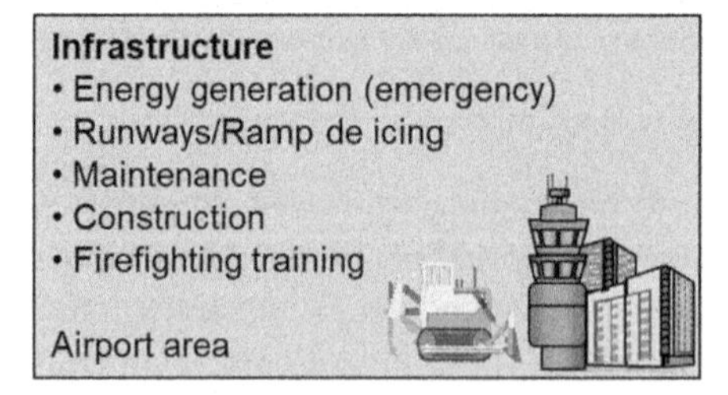

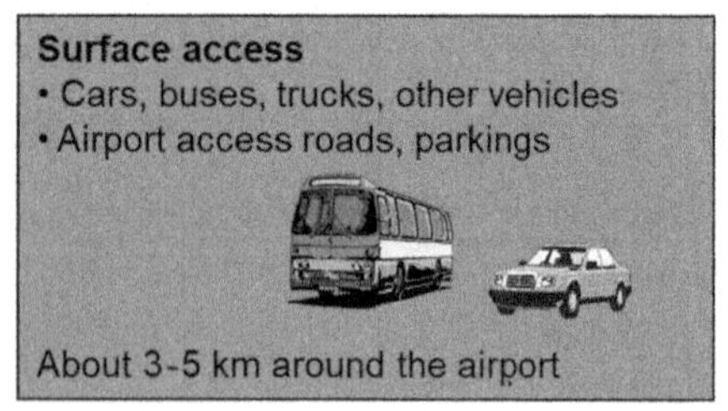

Fig. 9.1 Emission sources within the airport area.

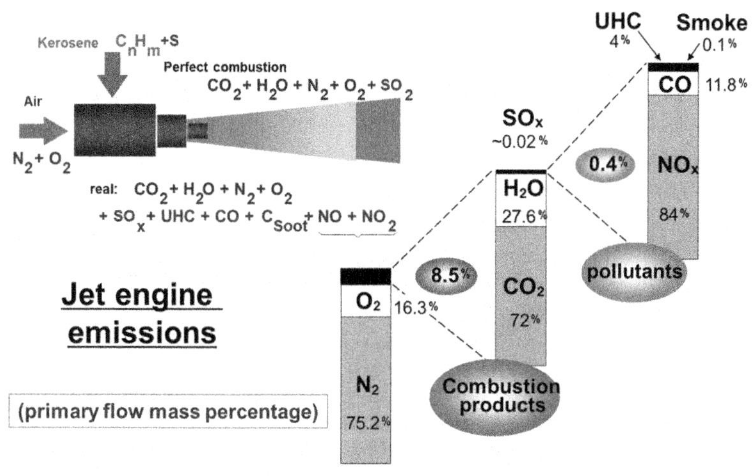

Fig. 9.2 Jet engine emissions (*IPCC*).

Table 9.1 **Pollutants regulated in Annex 16, Part 2**

Pollutant	Flight phase	Environmental effects
Unburnt hydrocarbons (HC)	Low power	Photochemical reaction (incomplete combustion), toxic, odor
Carbon monoxide (CO)	Low power	Toxic (incomplete combustion)
Nitrogen oxides (NOx)	High power	Photochemical reactions (smog), acid rain, toxic, ozone creation
Soot (C)	High power	Visibility, condensation trails (contrails)

As for noise, ICAO requires a certification of the pollutants emissions before granting the Type Certification to new aircraft types. The applicable regulation is given in Annex 16, Part 2. The list of the pollutants regulated by ICAO and their respective environmental effects is shown in Table 9.1.

The certification limits are the following:

- Smoke number limit is defined as the lowest of:
  - $SN \leq 83.6 \times (F_{00})^{-0.274}$
  - $SN \leq 50$
- For the other pollutants, the limits are based on the allowed emitted mass during the LTO cycle:
  - HC (g/LTO) $= 19.6 \times F_{00}$
  - CO (g/LTO) $= 118 \times F_{00}$
  - NOx (g/LTO) $= (40 + 2\ OPR) \times F_{00}$

where

F_{00} is the maximum certified engine thrust
OPR is the engine Overall Pressure ration

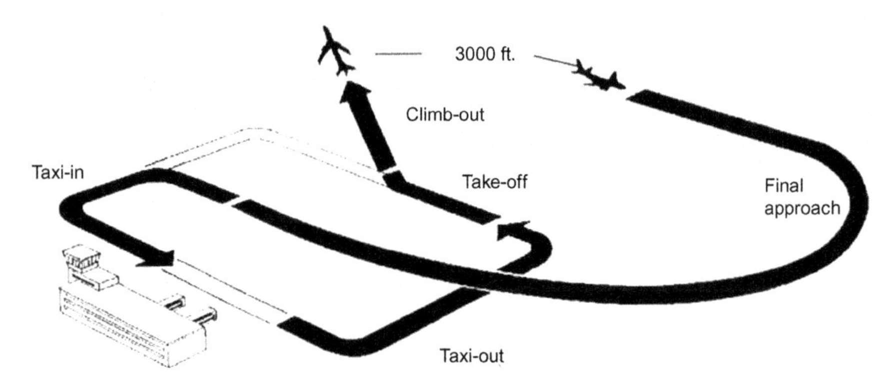

Fig. 9.3 LTO cycle.

Table 9.2 LTO cycle (Landing Take off), air operations under 3000 ft. (915 m)

Operating phase	Thrust	Time (min)
Take off	100% F_{00}	0.7
Climb	85% F_{00}	2.2
Approach	30% F_{00}	4.0
Taxi/ralenti	7% F_{00}	26.0

The landing and take-off (LTO) cycle includes the air operations under 3000 ft. (915 m), see Fig. 9.3, with the Take-off, Climb, Approach, and Taxi phases described in Table 9.2 in terms of engine thrust and duration.

LTO cycle emissions are basically dependent on the engine design. These emissions follow the ICAO Annex 16 Part II stringency evolution. Today, high Bypass ratio engines could have problems with NOx compliance. In 2004, ICAO approved CAEP/6 proposal to reduce 12% the limits for new certified engines after 01/01/2009. Later, in 2010, ICAO Assembly approved a further reduction of 15% for OPR $\geq$ 30 engines and 5%–15% for OPR < 30 engines, certified after 01/01/2013.

ICAO Environmental committee (CAEP) is preparing a future regulation on nonvolatile solid particles (nvPM), including visible (smoke) and nonvisible ones. In 2003, SAE emissions committee approved the measurement technical procedure for solid particles down to 10 nm. There is an existing engines test cell plan to calculate nonvisible weight and number solid particle emissions, using LTO cycle procedure, and determine the stringency levels for future regulation to be approved not later than 2019.

Local air quality in the airport area complies with general national standards. Some large airports may have little margins if ground sources are important. Pollutant dispersion in open space makes difficult to identify the source of each one. Theoretical models are still in very primitive phases. Air quality has been a limitative factor for some airport developments, as the third London Heathrow runway, still in discussion. The main problem would be NOx concentration.

Although fuel distilled from fossil fuel is the most characteristic nonrenewable material used by air transport, it is not the only one. Other nonrenewable materials consumed by the industry are:

- Some scarce metals, in particular, Titanium
- Some substances still in use although their production is already forbidden, like some CFCs, in particular Halon.

The fossil crude oil distillation produces the following substances (% in density):

- Kerosene and aviation gasoline (7%)
- Automotive gasoline (43%)
- Diesel and heavy fuel (23%)
- Lubricant oil and asphalts (16%)
- Waste (11%)

The most used types of aviation fuels are:

- High-octane gasoline (Grade 100, alternative engine)
- Jet A (Kerosene, mostly used in the United States)
- Jet A1 (similar to Jet A, with a—50°C freezing point, used by almost all international operators)
- Jet B (Wide-cut, less used, lower density than Jet A)
- JP (Wider military specifications)

Fuel consumption has a higher or lower impact on airlines operating costs, depending on oil price. This consumption depends on multiple technical and operating factors. The climate change eruption has made it a high priority.

Natural phenomena have always contributed to climate change. Climate changes prior to the Industrial Revolution in the 1700s can be explained by natural causes, such as changes in solar energy, volcanic eruptions, and natural changes in greenhouse gas (GHG) concentrations. Recent climate changes, however, cannot be explained by natural causes alone. Anthropogenic climate change is defined as "A climate change directly or indirectly chargeable to human actions that modify the composition of the global atmosphere added to the natural climate variability observed during comparable time periods" (UNFCC). Human activities that contribute to climate change are basically industry, agriculture, and transport.

The combustion of hydrocarbon is the chief source of man-made CO_2, which is a significant GHG. Other products of the combustion have also environmental effects and an indirect contribution to the climate change, but they are not regulated in Annex 16 Part 2 (Table 9.3). ICAO is studying nvPM certification in this Annex 16 Part 2

Table 9.3 Substances unregulated in Annex 16, Part 2

Substance	Flight phase	Environmental effects
Sulfur oxides (SOx)	Whole flight	Photochemical reactions, odor, greenhouse effect
Carbon dioxide (CO_2)	Whole flight	Greenhouse gas
Nitrogen oxides (NOx) out of LTO cycle	Climb/cruise/descent	Photochemical reactions, ozone creation, methane depletion
Water vapor (H_2O)	Whole flight	Condensation trails (contrails), cirrus clouds formation

and to create a Part 3 regulating the maximum levels of CO_2 emissions for new civil aircraft types (see Chapter 10). The selected metric is:

$$(1/SAR)_{AVG}/RGF^{0,24}$$

where $(1/SAR)_{AVG}$ is the specific air range average of three aircraft weights, representing the situation at the beginning, in the middle and at the end of a typical cruise. RGF is a factor representing the floor surface devoted to carry payload. The certification procedures were approved by ICAO General Assembly in 2013.

The maximum levels and the application date were established in the 2016 ICAO Assembly.

The Intergovernmental Panel on Climate Change (IPCC) is the international body for assessing the science related to climate change. IPCC assessments provide a scientific basis for governments at all levels to develop climate-related policies, and they underlie negotiations at the UN Climate Conference—the United Nations Framework Convention on Climate Change (UNFCCC).

ICAO asked in 1997 for an IPCC Special Report (1999) that was a compendium of all the available knowledge at that time. It contains:

- Atmospheric sciences
- Aeronautic technology
- Socioeconomic effects and reduction possibilities
- Scenarios of the possible aviation effects until 2050

Basic consequences of climate change are that Earth atmosphere and oceans are warming; snow and ice are diminishing, especially in Arctic areas; sea level is rising from 1.5 to 3 mm per year; each of the most recent decades has been successively warmer than any preceding decade since 1850. According to IPCC, increase in global average temperature should not exceed 2°C above preindustrial level, in order to prevent a catastrophic evolution. IPCC calls therefore for urgent and more decisive action, like the Paris Treaty (2015) ratification.

To explain climate change, the concept of Radiative Forcing is introduced. The difference between radiant energy received by the Earth and the energy radiated back to space is called Radiative Forcing (positive forcing warms the system). The effect of human activities on the Radiant energy balance is called anthropogenic radiative forcing. The aviation contribution to climate change is based on the fact that CO_2 is a GHG. NOx act as ozone precursors with a heating effect but also as methane destroyers, with a cooling effect. Soot and water vapor have a small heating effect, while sulfur oxides have a small cooling effect. Finally, water vapor may create condensation trails, known as contrails, and/or cirrus type clouds. This effect is not well known yet, but might be potentially dangerous from the climate point of view. The Radiative Forcing of the aircraft emissions as calculated in the TRADEOFF study (TRADEOFF, 2004) is shown in Fig. 9.4. The very different levels of scientific understanding of the impact of each substance on climate change can be appreciated.

The results of IPCC have been reviewed every 5 years. The latest (Fifth) Assessment Report was published in 2013, with the following findings for the air transport industry:

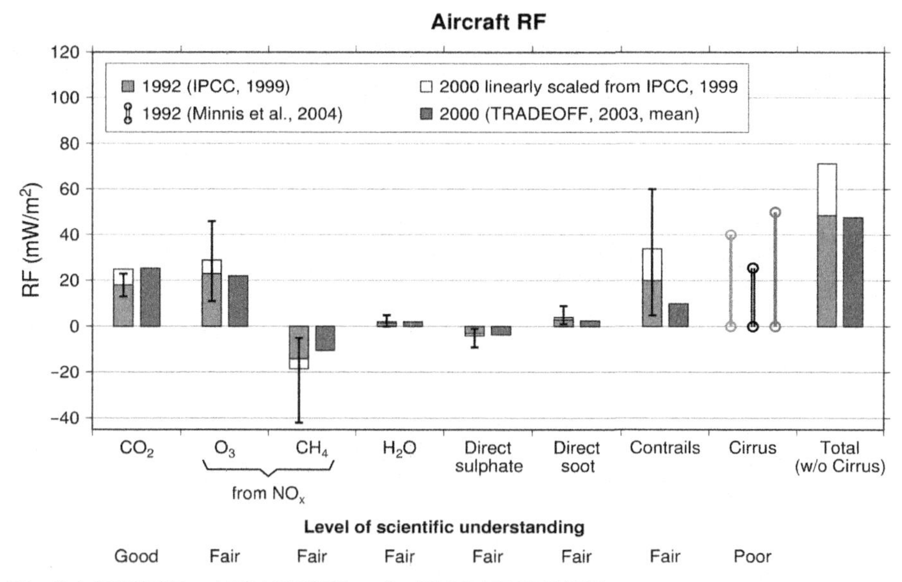

Fig. 9.4 IPCC RF and TRADEOFF study (*IPCC, TRADEOFF*).

- Aviation contributes about 4% to the anthropogenic greenhouse effect.
- Slightly over half of that contribution (about 2%–2.5%) corresponds to CO_2 emissions, a well-known effect.
- The rest is shared by indirect NOx effects, water vapor, and particles as ozone precursors, condensation trails, and cirrus clouds. Their effects are not totally well known and their quantification is not accurate.

The IPCC is currently in its Sixth Assessment cycle. During this cycle, the Panel will produce three Special Reports, a Methodology Report on national GHG inventories, and the Sixth Assessment Report (AR6). The AR6 Synthesis Report would be finalized in 2022 in time for the first UNFCCC global stocktaking when countries will review progress toward their goal of keeping global warming to well below 2°C while pursuing efforts to limit it to 1.5°C.

The different environmental impacts of air transport are not independent from each other. Sometimes, the reduction of one of the effects to be controlled may have opposed results on other environmental factors. The most frequent opposite effects are:

- Noise and fuel consumption (CO_2 emissions)
- NOx and HC
- NOx and fuel consumption (CO_2 emissions)
- NOx and nvPM

A classic example of these interdependencies was the Airbus A380 engines selection. Airbus 380 model was offered with two engine types: Rolls-Royce Trent 900 and Engine Alliance GP7200. The launching customer, Singapore Airlines, had selected the Trent engine and demanded the fulfillment of Heathrow night noise rules (category QC2). The solution for noise reduction was a larger diameter fan, a solution that also reduces the fuel specific consumption, but additionally adds more weight and more drag, worsen the fuel consumption in long-range flights. As a consequence, aircraft cruise drag increases and fuel consumption (and CO_2 emissions) went up around 0.3%–0.5%.

Design changes in engines combustion chambers are used by engine manufacturers to reduce the NOx emissions. These reductions in NOx emissions may result in an increase of other emissions. That was the case of the new combustion chamber that Pratt&Whitney developed for their PW4000 engine, under the TALON (Technology for Advanced Low NOx) program. The new combustion chamber made a substantial reduction in NOx emissions, as well as in CO and UHC levels, but increased largely the Soot level (although under the CAEP limits).

The replacement in the CFM56 of the single annular combustor by a double annular combustor is another classical example of interdependencies. The change produced the desired effect of reducing NOx emissions, and simultaneously Soot levels, but increased UHC and CO emissions, again still within CAEP limits.

9.2 Organizations with relevant activity in the air transport environmental regulation

There are a number of national and international organizations that play a fundamental role in the regulation and management of air transport environmental issues. They are listed in Table 9.4 and their activities briefly described in this section.

International Civil Aviation Organization (ICAO)

ICAO is the United Nations Agency in charge of international civil aviation. It was created in 1944, according to the Chicago Convention Part II. With 191 Member States its mission is to ensure a safe, reliable, and cost-effective world civil aviation

Table 9.4 Organizations with relevant activity in the air transport environmental regulation

World	Administration	ICAO
		UNFCCC
	Private entities	IATA
		ACI
		ICCAIA
		CANSO
		ICSA
Europe	Administration	EU
		CEAC
		EUROCONTROL
	Private entities	AEA
Other regions		FAA
		ATAG

development. It has a mandate to establish and keep updated the Standards and Recommended Procedures (SARPs) in the 19 Technical Annexes to the Chicago Convention and the adoption of procedures and guidance material.

The following is a list of the Standards and Recommended Procedures (SARPs) related to environment protection:

- Annex 16, Environmental Protection
  - Part 1 Aircraft Noise
  - Part 2 Aviation Engine Emissions
  - Part 3 (in discussion) CO_2 Emissions
  - Part 4 (in discussion) CORSIA
- Environmental Technical Manual (Doc 9501)
- Airport Planning Manual (Doc 9184) Part 2—Land use and environmental control
- Circulars
  - (218) Economic implications of future noise restrictions on subsonic jet aircraft
  - (303) Operational opportunities to minimize fuel use and reduce emissions (to be replaced by a Technical Manual)
- Guidance Material:
  - ICAO's policies on charges for airports and air navigation services (Doc 9082)
  - Guidance on the Balanced Approach to aircraft noise management (Doc 9829)
  - Guidance on aircraft emission charges related to local air quality (Doc 9884)
  - Draft guidance on the use of emissions trading for aviation (Doc 9885)
  - Airport air quality manual (Doc 9889)
  - Recommended method for computing noise contours around airports (Doc 9911)
  - Environmental Management system (EMS). Practices in the aviation sector (Doc 9968)

In addition to the SARPs, ICAO issues Policies and Procedures, which are consolidated statements of continuing ICAO policies and practices related to environmental protection. For instance, the following Assembly resolutions are of particular importance:

- Assembly Resolution A39-1 (2016)—General provisions, Noise and local air quality (Appendix A to H)
- Assembly Resolution A39-2 (2016)—Climate change (includes an Annex on MBM conditions)
- Assembly Resolution A39-3 (2016)—Global Market-based Measure (MBM) scheme

The purpose of those resolutions is:

- To limit or reduce the number of people severely affected by noise
- To limit or reduce aviation emissions impact on local air quality
- To limit or reduce aviation GHG emissions contribution to climate change

Appendixes to the Resolution A39-1

A. General
B. Development of SARPs and/or Guidance Material relating to the quality of environment
C. Policies and programs based on a "balanced approach" to aircraft noise management
D. Phase-out of subsonic jet aircraft, which exceed the noise levels in Volume I of Annex 16
E. Local noise-related operating restrictions at airports
F. Land-use planning and management

G. Supersonic aircraft—The problem of Sonic Boom
H. Aviation impact on local air quality

Appendixes to the Resolution A39-2
Guiding principles for MBMs for international aviation

- Support aviation sustainable development
- Support mitigation of GHG emissions from international aviation
- Contribute toward global goals
- Transparent and administratively simple
- Cost effective
- Avoid duplications in CO_2 accounting
- Minimize market distortions
- Fair treatment of international aviation in relation to other sector
- Recognize aviation past achievements
- Not impose inappropriate economic burden on international aviation
- Appropriate access to all carbon markets
- Based on CO_2 reduction efficiency
- Include de minimis provisions
- Revenues to be applied internally
- Emissions reductions to be identified
- Consideration of CBDR, SCRC, and no discrimination principles

Committee on aircraft environmental protection (CAEP)

The mission of CAEP is to perform specific studies, approved by ICAO Council, on the control of aircraft noise and gaseous emissions from aviation engines, taking into account what is technically feasible, economically reasonable, and environmentally beneficial. It is composed of experts representing 23 States, and 12 Observers with voice but no vote. The CAEP Plenary meets once every 3 years, before the ICAO Assembly.

CAEP meetings and their most important decisions are summarized hereafter:

CAEP/9 (February 2013):

- Propose to the Assembly new noise standards (Annex 16 Chapter 14), to be applied starting in 2018/2020
- Prepare not later than 2016 certification requirements for nvPM (a standard to be approved by 2019)
- Propose to the Assembly certification requirements for commercial aircraft CO_2 emissions (a standard to be approved by 2016)
- Prepare a study for the 2016 General Assembly on the application of a worldwide Market-Based Measure system to control global aviation CO_2 emissions

All these proposals were approved during the 38th ICAO Assembly in September-October 2013.

CAEP/10 (February 2016):

- Review and analysis of the civil supersonic aircraft noise requirements
- Propose to the Assembly the certification requirements for nvPM (a standard to be approved by 2019)

- Propose to the assembly a CO_2 emissions standard for the certification of civil aircraft
- Propose to the Council the basic scheme of a worldwide Market-Based Measure system to control global aviation CO_2 emissions to be developed and approved in the September 2016 Assembly

The present CAEP goals for the different environmental impacts are:

- Noise:
  - Technology and Standards review
  - Analysis of existing experience on Balance Approach application. SST standards
- Emissions:
  - Technology and Standards review
  - Operating measures (Energetic efficiency)
- Interdependencies among different measures
- Market-based measures (MBM) and possible worldwide application

United Nations Framework Convention on Climate Change (UNFCCC)

UNFCCC is a convention that was signed during the Rio Summit in 1992, and entered into force in 1994. Its goal is to stabilize GHG atmospheric concentration at a level that avoids dangerous interference of human activities in the climate system. Its executive body is the Conference of the Parties (CoP), meeting once a year (latest ones in Paris in November 2015, Marrakech in November 2016).

Kyoto Protocol (1997)

It was approved in December 1997during COP/3 and entered into force in February 200, after Russia's ratification. The Developed States (listed in Annex 1) committed to reduce their global GHG emissions in the 2008–12 period at 95 of the 1990 levels. Each State had a different target, according to its per capita emissions. The European Union as a whole had an 8% reduction target to be distributed among its Member States (for instance, Spain's target was an increase of no >15%).

The control of international aviation and maritime traffic was left to the UN organizations ICAO and IMO, respectively.

Paris agreement (2015)

Approved in December 2015 during COP/21, it will enter into force when ratified by 55 States with at least 55% of the World greenhouse emissions. This is expected to happen at the Marrakech COP in 2016.

The Paris agreement requires the adoption of measures to keep global warming at 1.5°C average temperature and, in any case, <2°C. Although it does not refer directly to aviation, the possibility of using air transport as a levy-type means of payment for this policy cost is not excluded.

International Air Transport Association (IATA)

IATA was created in Havana in 1945 as a voluntary association for international airlines, with initially 260 members. It represents the largest association of international airlines with its members serving around 95% of the total international and 84% of the global world air traffic. IATA has six permanent committees: Financial, Operations, Industry Affairs, Cargo, Legal, and Environment.

The IATA Environmental Committee (ENCON) comprises 19 Member Airlines and meets twice a year. It also includes observers from regional airline associations, manufacturers, and fuel suppliers. Its main purposes are:

- Submit to the Board of Governors airlines common positions and proposals on environmental issues
- Support ICAO as the appropriate forum for developing global environmental standards and policies for air transport
- Assist members in managing their impact on the environment

Airports Council International (ACI)

ACI was created in Washington in 1948, with headquarters currently in Montreal since 1991. It has about 600 members operating over 1650 airports in 176 countries (96% of world traffic).

The ACI permanent Environmental Committee has 25 members. They cover not only noise and emissions but also ground, water, water, wildlife, and resource management. Its main goal is to ensure that traffic growth can be maintained within the environmental capacity limitations imposed.

Civil Aviation Navigation Services Organization (CANSO)

CANSO was created in 199 and has its headquarters in Hoofddorp (Holland). It comprises 158 ATM providers members (87 members covering over 85% of world air traffic movements plus 71 associated). The main goals are:

- Provide safe, efficient, and economic ATM services
- Optimize ATM routes and procedures in order to reduce noise and emissions

The International Coordinating Council of Aerospace Industries Associations (ICCAIA)

ICCAIA represents major aircraft, engines and aeronautical equipment, and aerospace industries associations (AIA, AIAC, ASD). They provide the technological base for long-term solutions and help to perform cost-benefit analysis of different environmental proposals. Their key role is the development of quieter, cleaner, and more fuel-efficient aircraft.

International Coalition for Sustainable Aviation (ICSA)

ICSA was created in 1999 with the purpose of participating in the CAEP work, representing the NGOs with a single voice. Its most active members are US and European organizations specialized in energy and transport, like the International Council for Clean Transport (ICCT) and Transport & environment (T&E).

ICSA main goals are:

- Noise and emissions reduction in aviation
- Develop strategies for improving aviation environmental impact
- Evaluate the potential impact of environmental regulation proposals

European Union (EU)

The European Union was legally established by the Rome Treaty in March 1957. It has 28 Member States (19 of them sharing the Euro currency). Some others are candidates to join. The EU main government bodies are the Commission, the Parliament, and the Council.

The environmental protection is one of the official goals of the EU. Some EU environmental regulations applicable to civil aviation are:

- Directive 1996/62/EC: On ambient air quality assessment and management
- Directive 1996/82/EC: On the control of major accidents hazards involving dangerous substances
- Directive 2002/30/EC: On the establishment of rules and procedures with regard to the introduction of noise-related restrictions at EU airports
- Directive 2002/49/EC: Assessment and management of environmental noise
- Directive 2003/87/EC: Emissions Trading (modified in 2008)
- Directive 2003/96/EC: Energy taxation
- Directive 2008/101/EC: Modifying 2003/87 to include aviation activities in the European Emissions Trade System, starting in 2012

The European Civil Aviation Conference (CEAC)

CEAC was created in 1955 and in headquartered in Paris. 44 European Member States are represented. CEAC harmonizes policies and procedures in the European aviation sector, developing technical standards through the Joint airworthiness Authorities (JAA) and taking care of environmental aspects in the ANCAT group (Abatement of Nuisances caused by Air Transport. The working groups of ANCAT are:

- AIRMOD: Aircraft Noise Modeling
- PLANO: Operational Noise Mitigation procedures around airports
- EMTRA: Emissions Trading
- ERLIG: Emissions-Related Landing charges Investigation Group
- TANC: Transport aircraft Noise Classification
- Environmental indicators

EUROCONTROL (The European Organization for the safety of Air Navigation)

EUROCONTROL was created in 1960, with headquarters in Brussels. It has 42 members (41 European States and the European Union). EUROCONTROL harmonizes

European Air Navigation Services, receives users' payments, and distributes them among Member States.

EUROCONTROL cooperates in the development of operating procedures for reducing noise and emissions. It also keeps statistics of all flights within the European air space.

Airlines for Europe (A4E)

Group of European airlines that has replaced in 2016 to the former AEA. AEA was established in 1952 in Brussels. It had 22 member European airlines and it is considered the preferred contact to represent airlines in front of European Commission and Parliament. AEA had an Environmental Committee. While AEA was integrated only by scheduled airlines, its replacement, A4E, has every type of airlines, among them, the three largest groups of schedule airlines (AF-KLM, IAG, Lufthansa) and the three largest low-cost operators (EasyJet, Norwegian, and Ryanair).

FAA (Federal Aviation Agency)

FAA is the organization of the US Department of Transport in charge of civil aviation technical issues. It has an Environmental Department coordinating the certification rules in FAR 33 (engine emissions) and FAR 36 (aircraft noise). FAA leads the US environmental aeronautic research, in collaboration with the Environmental Protection Agency (EPA).

Air Transport Action Group (ATAG)

ATAG was created in 1991 and has its headquarters in Geneva. ATAG is a global coalition representing all sectors of the air transport industry: airlines, airports, manufacturers, and air navigation service providers. There are other members from different sectors: tourism institutions, chambers of commerce, travel agencies, trade unions, etc. ATG purpose is to promote the aviation sustainable development. It publishes a number of environmental documents.

9.3 Environmental impact reduction

The environmental impact reduction goal is, in the long-term goal, making air transport growth compatible with its sustainability, having as partial goals the reduction of the number of people seriously affected by aeronautic noise, keeping the air quality in the airport zone within the regulatory limits and achieving air transport growth without increasing its effects on climate change.

While all the civil aviation stakeholders share the environmental impact reduction targets, they do not agree on the price to be paid and who has to do it, leading to a conflict of interest:

- Airlines ask for no-growth limits and the adoption of measures not endangering their economic results
- Airports propose giving priority to operating restrictions

- Manufacturers wish more research and an accelerated fleet renewal
- NGOs support airline offer reduction and ticket price increases

This conflict of interest led to the adoption of the Balanced Approach to tackle airport noise problems in the most effective way, and similarly to the Four Pillars to deal with aviation emissions reduction. Historically, aviation emissions reduction had been concentrated on improving local air quality. The recognition of climate change importance changed this perspective and made it clear that the certification approach, like Annex 16 Part 2 rules, was not sufficient to manage the problem. In 2004 AEA presented a formulation to cover both aspects with a set of measures, following the Balanced Approach line, which, with minor modifications, reached general consensus.

Another important aspect to be considered is that each one of the environmental impacts (noise, NOx, unburnt HC, smoke, and CO_2 emissions) cannot be independently reduced without affecting some of the others, because there are interdependencies. Measures must be evaluated taking into account their global effects, not only in terms of one of the impacts reduction. Some of the relationships among the different impacts are still in the research phase.

AEA launched the so-called Four Pillars policy in 2003, as an alternative to the proposal of imposing taxes on emission levels. The Four Pillars policy suggests the need of actions in four different areas:

- Research and development of better technology
- Infrastructure improvements (airports and air navigation services)
- Optimization of operating procedures
- Introduction of MBM

The first three "pillars" have been addressed already in Chapters 4, 8, and 6, respectively. The MBMs will be reviewed next.

9.4 Market-based measures

Environmental impact management, in addition to the solutions based in the technology, may use different economic actions, most of them in the shape of MBM. The most promising options, explored by ICAO, are:

- Voluntary agreements between the Administration, some operators, the consumers, or any combination of those three groups
- Taxes imposed by Central, Regional, or Local Administrations
- Charges related with different aeronautical payments, such as airport, air navigation, or passenger fees
- Emissions Trading System (ETS) as defined in Kyoto Protocol

The goals of this type of economic measures are:

- To complement traditional regulatory measures with others with participation of stakeholders other than the Administrations
- To offer flexibility to the operators looking for the most economic option to comply with the rules

- To create incentives for operators and/or consumers with the purpose of modifying their behavior
- To "internalize" the "externalities," building up prices, which include the total cost of the activity

Domestic air transport is included in the national GHG inventories, regulated in the Kyoto Protocol (1997). The Protocol gives ICAO a mandate to regulate GHG emissions produced by international air transport (3.5%–4.0% of total anthropogenic effects, 2.0%–2.5% of CO_2 emissions). ICAO analysis shows that a mix of short-term voluntary agreements and long-term emissions trading is the most efficient MBM procedure for emissions limitation. ICAO recommends the ETS application through its inclusion in bilateral or multilateral traffic agreements. MBM were evaluated by CAEP in the 2001–04 period, and the results showed the following conclusions:

- Voluntary Agreements are beneficial in the short term, but are not a solution in the long term
- Taxes show very low cost/benefit efficiency
- Charges may be acceptable for local problems, but act like taxes in a global application
- Emissions Trading is the most efficient system, when applied in open regime, i.e., allowing trading with other industrial sector.

Most of the economic measures for environmental management are based on the ideas of Pigou and Coase, both 20th century economists. In both cases, the problem appears from working in areas where the market is not efficient in the allocation of prices and resources. The major difference among them is their respective visions on the Administration role in these cases. That is a vital element for Air Transport, because its operation produces a great externalities volume.

Arthur C. Pigou (1877–1959) was the author of the distinction between private and social marginal costs. He defined externalities as costs not supported by the user, but inflicted on the rest of the society. He suggested that Governments can correct those market deficiencies through a mix of taxes and subsidies, in such a way that "externalities are internalized." These theories were published in his book *Wealth & Welfare* in 1920.

Ronald H. Coase (1910–2013) however considers that the externalities internalization does not need the Government intervention. Stakeholders might arrive to particular agreements, with a final result very similar to the one obtained by the application of Pigou's ideas if there is no transaction cost *(The Coase Theorem).* The Administration role is reduced to facilitate the negotiation among private parties and control the general rules. His book, *The problem of social cost,* was published in 1960.

An external cost, or "externality," is a cost that comes from an economic activity, is paid by someone out of the market transaction and, therefore, is not included in the market price. As a result: overproduction and excessive consumption of products economically inferior and underproduction and scarce consumption of products economically superior.

External costs may be considered from the point of view of compensation of resulting damages or investment in prevention measures.

The different external costs schemes can be:

- Producer—producer: a company A pollutes a river the water of which is used by another company B located downstream

- Consumer—consumer: a person makes a barbecue in his garden and the smoke and odor reaches his neighbors' house
- Producer—consumer: aircraft noise disturbs the sleep of people living in houses close to the airport

In theory, economic measures are used to optimize the production and consumption sequence through the application of:

- Treasure taxes
- Operator incentives (positive or negative, with or without recaudatory purposes)
- Financing compensatory mechanisms (fiscal exemptions)
- Investment incentives
- Traded allowances

They are generally based on the *Polluter Pays* Principle, established in paragraph 16 of the Rio Summit final Declaration (1992). In practice, however, the use of economic measures in commercial aviation is supported from other points of view:

- Fiscal leveled playground
  - *All fossil fuels should be taxed*
- Financing of Governmental institutions
  - *Customs, infrastructures, security*
- Environment protection
  - *Decrease air transport demand to reduce emissions*
- Make a better world
  - *Development help for Third World countries, like the Chirac tax for Africa in France*

The first type of MBM is voluntary agreements. There may be different types of voluntary agreements, for instance:

- Between the Administration and the industry, with the commitment to adopt energetic efficiency measures and good practices. Examples can be found in Canada and Japan. ICAO approved recommendations for this type of projects in 2004.
- Between some industry sectors and the public opinion. IATA has settled energy efficiency targets and publishes yearly the achieved position versus those targets. One of these agreements established a total efficiency improvement for the period 1990–2012 of 26%, resulting in an annual efficiency improvement for that period of 1.1% (Fig. 9.5).
- Between airlines and their customers, who are offered the possibility to pay an extra amount of money to offset the CO_2 emitted by their part of the flight.

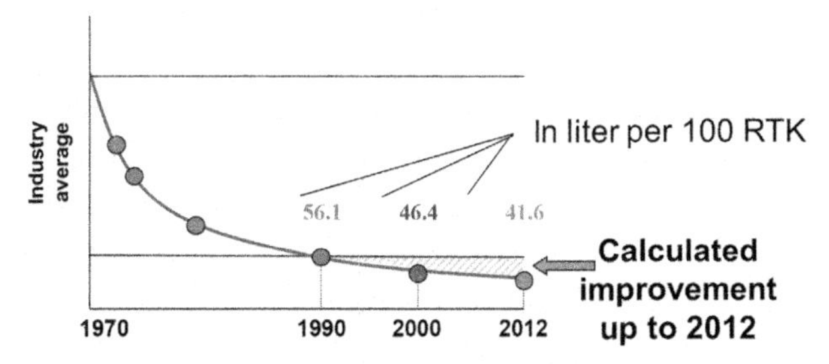

Fig. 9.5 IATA voluntary agreement (*IATA*).

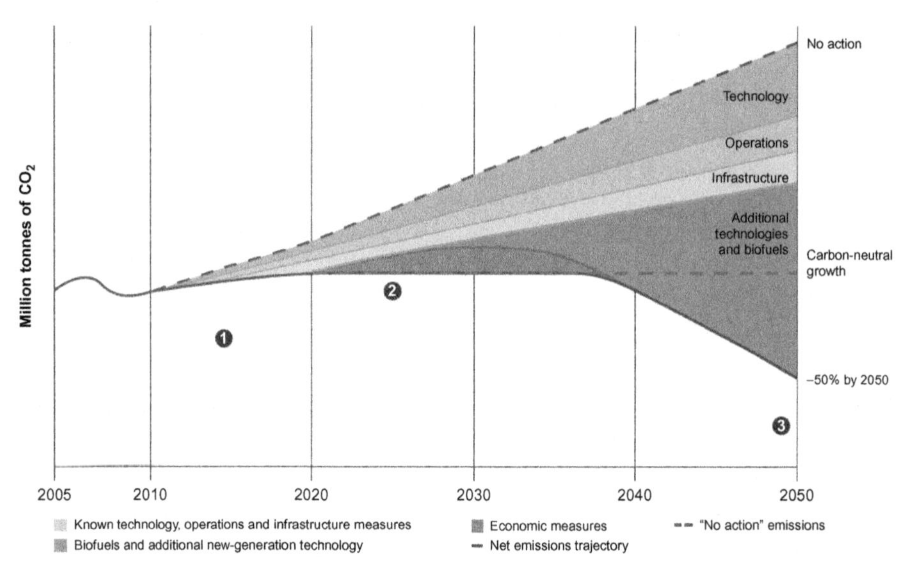

Fig. 9.6 CO_2 emissions future development.

The current IATA goals were adopted by the IATA Council in 2009. The purpose is to achieve a traffic growth without increasing CO_2 emissions (Carbon-Neutral Growth or CNG) from 2020 levels (using MBM as well). Between 2020 and 2050, commercial aviation would reduce CO_2 emissions 50% with respect to 2020 (Fig. 9.6). IATA has developed an energy auditing program and makes active best practice diffusion among its members.

Taxes are the second type of MBM. Taxes may adopt two different lines: fuel (or CO_2 tax), or a general environmental tax applied to the passenger.

Concerning fuel taxes, the Chicago Convention in its Article 24 (and an ICAO Council Resolution on 14/12/1993) recommends Member States not to apply taxes on the products uploaded in an international flight aimed at being consumed during the flight. Obviously, fuel is one of these products. This is confirmed in the ICAO Doc 8632 on charges and taxes and it is included in the traffic rights bilateral agreements between States. Many bilateral agreements on traffic rights used to have a reference to that Article. EU considers this provision inadequate and defends its suppression. The EU regulation allows the introduction of this fuel tax inflight between two countries if both States agree on this point. There have been no application so far. Outside the EU, some States apply fuel taxes to domestic flights (USA, Japan, Norway) but with much lower levels than to the road transport fuels.

Regarding environmental taxes, some States like Austria and Germany have taxes with the revenues assigned to environmental programs. The United Kingdom collects the Air Passenger Duty (APD) from the passengers boarding in UK airports. APD gets about 4500 MUS$ yearly. Up to now, the experience from this approach is negative in terms of economic benefit per unit of cost, with a cost close to 1000 € per CO_2 ton reduction.

Taxes and charges have very different definitions in ICAO texts. Taxes are paid to collect funds for local or national Governments and may be used for any purpose.

Charges are paid to compensate the cost of different services provided to the operators. They are finalists. ICAO defends the use of charges rather than taxes. Both elements are put together under the term levies. The questions are: is the concept applicable to environmental charges? What services are they supporting?

Noise charges are being used for the last 40 years. There are 28 States using them in different ways. Normally, they have different values in each airport, generally related to the landing charge as a surcharge. Collected money is paid for noise control and monitoring, house insulation, zoning, and any other noise-related airport activity. It may be based on the certified noise values of each aircraft type or in the actual noise registered by the airport-monitoring system. ICAO recommends those schemes to be in compliance with a number of recommendations given in Doc 9082, Chapter 30:

- To be applied only in airports with a noise problem
- The collected amount of money not to be higher than the cost of the airport noise reduction programs
- Preferably associated to the landing charges
- Taking into account the Annex 16, Part 1, certificated values
- Do not discriminate among operators or aircraft types
- Not being so expensive as making uneconomic the operation of certain aircraft types

ICAO has established its position on Emissions charges in three documents. Existing schemes comply with that guidance material. The documents are:

- Doc. 9082/7 referring to airport and air navigation charges policies
- Council Resolution (09/12/1996) on aeronautical charges and taxes
- Resolution A35-5 of ICAO 2004 General Assembly

Doc. 9082/7 confirms the application to emissions charges of the same general guidance than for noise charges. They are the following:

- They will be based on the cost of mitigating aircraft engine emissions impact, as far as those costs can be clearly identified and directly attributed to the air transport.
- In addition to those costs, the system must be neutral in total revenues.
- There will be no discrimination among the users.
- Charges will be transparent and settled in consultation with the operators.

The 1996 Council Resolution accepts the principle of each economic sector paying for the complete cost of its environmental impact and recommends that the money collected through emissions-related charges is spent in mitigation actions to reduce the environmental impact of those emissions, defending the nondiscrimination of air transport with respect to other transportation modes. The 2004 General Assembly affirms that direct CO_2 charges are not adequate to mitigate climate change and declares its preference for the use of an emissions-trading mechanism.

Emissions charges are being used since 1997, only for NOx. Today there are landing charges modulated with NOx emissions in the largest 12 airports in Sweden, Basel, Berna, Geneva, Lugano, and Zurich airports in Switzerland; Gatwick, Heathrow, and Luton airports in the UK; Dusseldorf, Frankfurt, Hamburg, and Munich airports in Germany; and Copenhagen airport in Denmark.

All of them use the ERLIG system, based on certified values and proposed by ECAC, with little local variations. The ECAC recommendation 27-4, prepared by the ERLIG Group, proposes a continuous scale, based on the LTO-cycle-certified emissions. ERLIG divides aircraft considering the margin of its engine-certified NOx values with respect to the Annex 16, Part 2, levels. Some old engines have good NOx margins but very poor figures in HC and CO. The scale is applicable if HC values do not reach a certain maximum value. In that case, the NOx surcharge would be the highest, i.e., over a top level of HC emissions, the aircraft is classified in the most expensive group, independent of its NOx values.

The application so far of taxes and charges shows some of the internalization difficulties:

- How can the environmental damages be evaluated? A number of values like annoyance, life quality loss, health effects, etc., have to be put in monetary terms.
- Who is to be blamed for the damages? Decide whether the airport, the airline, the air transport users, the Administration have individual or shared responsibilities.
- How can environmental impacts be forecasted? If a prevention system is adopted, take a decision on to what extent the damaging factors should be prevented.

And raise further questions on the internalization policies:

- Should they cover environmental damages only or safety, congestion, and other externalities as well? This may produce large differences depending on the transportation mode.
- Should they count individual effects (noise, CO_2) or all together? It might give advantage to some transport companies with respect to others.
- Should they apply to aviation or to all the transportation modes? Road transport has many operators and is more difficult to regulate than the other modes, where public transport is dominant.

When technical measures and improved operating procedures are not enough, the dilemma is to implement regulatory or economic measures to complement them. In the absence of a Chicago Convention modification, it is not possible to apply generalized taxes on the international flight kerosene and the States are not willing to include it in bilateral agreements.

According to ICAO, the most environment-efficient measure is Emissions Trading, one of the three flexibility mechanisms included in the Kyoto Protocol, in addition to Joint Implementation (JI) and the Clean Development Mechanism (CDM).

Emissions Trading is a cap and trade system. It sets up a cap for a participant emissions and provides them allowances to emit, i.e., rights to emit a fixed amount of emissions per year. The system allows participant to buy and sell emissions allowances, i.e., rights to emit a fixed amount of emissions per year. This way, emission goals to be met can be established in the most cost-effective way by letting the market determine the lowest-cost pollution abatement opportunities. At the end of each trading year, participants have to hand over or surrender allowances corresponding to their actual emissions in that year. Participants can either sell unused allowances to other participants in the scheme or must buy the allowances needed to cover their extra emissions from other participants on the open market. The goal is to minimize the cost of reducing emissions globally.

From the kerosene combustion chemical reaction:

$$C_{12}H_{23} + 17.75O_2 \rightarrow 12CO_2 + 11.5H_2O$$

It can be determined that 1 kg of kerosene combustion produces about 3.15–3.16 kg of CO2 with small differences depending on the fuel specification. Until the scientific community can offer a better knowledge of the effects of all other factors on the climate change, most of the technical and economical efforts are concentrated on CO_2 emissions.

The European Emissions Trading System (EU ETS) has been applied in 6 economic sectors (46% of EU CO_2 emissions) since 2005. In 2007 EU Commission published a draft Directive to include civil aviation in the EU ETS, starting 1st of January 2012. The proposal became the Directive 2008/101/EC, issued in January 2009. Different than Kyoto, it is based on a nonnational scheme, with the Airlines as the stakeholders and owners of emission allowances. It does not distinguish between domestic and international flights.

The Directive covers only CO_2 emissions from civil aviation flights (about 2.0%–2.5% of world total). Reference level is the annual average emissions of the 2004–06 period calculated, according to EUROCONTROL data, in 219 Mton (slightly over 1/3 of the world aviation total for those years). In 2012, there were allowances for 97% of the reference level. In 2013, it was down to 95%, and will be constant until 2020. 82% are free permits, distributed by *benchmarking,* according to the 2010 traffic, measured in Revenue Ton-Kilometer (RTK). In addition, 15% has been auctioned and 3% is reserved for new Airlines, not operating in 2010 but with flights in 2012, and those having a high rate of growth (≥18% annual). System monitoring, possible sanctions, and the payment collection are kept at national level. Collected money will cover administrative expenses and programs for climate change mitigation.

International aviation emission allowances are not included in the Kyoto Protocol national inventory and cannot be used by other economic sectors, but airlines are authorized to buy emission allowances from other activities as is determined in the Protocol. ETS covers all flights by airlines and private operators with aircraft with a MTOW higher than 5700 kg. Each EU Member State controls its nationality Airlines and those from non-EU States emitting more CO_2 in the flights to this country than in those to other EU States Operators' information is under a Monitoring, Reporting, and Verification (MRV) obligation, in the terms of a Commission decision, approved in June 2009. It would include all flights in and off EU airports since 1st of January 2012. The system passed a general test in 2010, and used also for the initial distribution of emission allowances.

The efficiency unit is the amount of emitted CO_2 per RTK, adding up passengers, freight, and mail, including APU emissions. The weight of the emitted CO_2 is computed multiplying the consumed fuel weight by 3.15 (emission factor), when it is standard kerosene. Biofuels have zero emission factor. Payload weight is thc onc stated in each flight mass & cg documentation or, by default, 100 kg per passenger and his baggage. Flight distance is the orthodromic plus 95 km.

The extraterritorial application (inclusion of non-EU airline flights) does not comply with ICAO recommendations (legal and diplomatic problems with USA and other countries, menacing commercial retaliation. Three US airlines brought UK Government to the Court in November 2009, but the European Court of Justice failed against them). It does not exclude countries with <1% of total international RTK, established in the ICAO Resolution A37-19, but adopts a simplified emission calculation procedure for operators with few flights. There are some special exclusions (ultraperipheric territories as Canary Islands, small aircraft MTOW≤5700 kg, low number of flights airlines, Services of Public Interest). Each operator MRV is assumed by its national EU State or the Member State having the most emissions, according to EUROCONTROL data from a Company if the operator nationality is not of an EU State. The operator lists attributed to each Member State are published by the European Commission. MRV and sanctions procedures must be audited by recognized organizations, and approved by National Authorities.

The system is open to new countries. In December 2009, Iceland, Lichtenstein, and Norway joined. In 2013 Croatia was added, as a new UE Member and Switzerland is presently discussing its adhesion.

The growing opposition of some important States, like Brazil, China, Russia, and USA, has brought back the ETS discussion to ICAO. In November 2012, EU decided *"stop the clock"* on the inclusion of flights coming from or going to non-EU countries in the European ETS, waiting for a new global system to be approved by ICAO. ETS rules remain the same for internal EU flights. In September 2013, ICAO Assembly ordered a study on a worldwide MBM to be approved in the 2016 Assembly and applied in 2020.

World airlines increase their fuel efficiency 1.5%–2.0% yearly. If the served demand increases 4.0%–5.0% annually, air transport sector will be an emission allowances net buyer forever. A closed ETS (only commerce among airlines) will be very expensive due to the small number of sellers. An open ETS would be cheaper for aviation, although general price levels might increase. Present economy weakness has reduced the allowance price down to 4€ per CO_2 ton. It remains the legal problem of mixing Kyoto domestic flights allowances with the new "aviation" ones, not included in the Protocol.

In June 2009 Decision on Monitoring, Reporting and Verification (MRV) was approved. Operators presented their MRV plans before the end of 2009 and put them into practice in 2010. Based on those data, the Commission distributed allowances on February 2012, but all the system had to be corrected to exclude external flights after the "*stop the clock*" decision. First payments were made by April 2013, but some airlines (mainly from China and India) have rejected to pay by indication of their Governments. Small operators, most of them noncommercial, may use a simplified mechanism, with emissions calculated by EUROCONTROL to avoid MRV complexities.

Following the 2013 Assembly decision, two ICAO working groups started to work in March 2014. The *Environment Advisory Group* (EAG, 17 States, of politic character) tries to reach a consensus on the main features of the worldwide MBM to be proposed to 2016 Assembly. The *Global Market-based Measure Technical Task Force* (GMTF, inside CAEP, of technical character) makes the calculation of different alternatives and develops MRV procedures and legal conditions that might be needed.

In August 2016, the option with most supporters appears to be an offsetting system taking 2020 emissions as a baseline.

In October 2007, ICAO Assembly approved a guidance to create a worldwide ETS but a great majority of States rejected its immediate implementation. The core problem is that Kyoto Protocol is only applicable to industrialized countries and not to every State, while the proposed ETS would cover all countries. To solve this problem,

ICAO set up GIACC (Group on International Aviation and Climate Change). Its recommendations were published in May 2010, approved by ICAO 37th Assembly in October, and reviewed in the 2013 38th Assembly.

The resolution A38-18 on Climate Change established a collective annual goal of improving 2% the energetic efficiency (measured in fuel per RTK) in the 2011–20 period, with a possible extension up to 2050. Member States must send ICAO their Action Plans by June 2015, and then every 3 years. Those states introducing MBMs must follow ICAO recommendations. The flights to and from States whose international traffic do not reach 1% of the total world international RTK may be exempted of MBM application (see Table 9.5). CAEP will study a CO_2 certification standard and a tentative worldwide MBM scheme for 2020 enforcement (see Chapter 10).

Table 9.5 RTKs per country

State	RTK(2009) $\times 10^6$	% of total	RTK(2008) $\times 10^6$	% of total	Δ% 09/08
1.-USA	54,372	15.1	58,958	15.4	-8
China	28,773	8.0	30,813	8.1	−7
Germany	26,243	7.3	29,156	7.6	−10
United Kingdom	22,782	6.3	23,378	6.1	−3
U. Arab Emirates	21,822	6.1	19,337	5.1	13
France	17,178	4.8	18,996	5.0	−10
South Korea	15,589	4.3	15,753	4.1	−1
Holland	13,111	3.7	14,305	3.7	−8
Singapore	12,973	3.6	15,902	4.2	−18
10.-Japan	12,665	3.5	14,354	3.8	−12
Ireland	8008	2.2	7291	1.9	10
Canada	6942	1.9	7137	1.9	−3
Australia	6923	1.9	7633	2.0	−9
Thailand	6539	1.8	7112	1.9	−8
Spain	6361	1.8	6849	1.8	−7
Qatar	5621	1.6	4922	1.3	14
Malaysia	5251	1.5	5883	1.5	−11
Russia	5168	1.4	5551	1.5	−7
India	5086	1.4	4932	1.3	3
20.-Turkey	4855	1.4	3924	1.0	24
Luxembourg	4688	1.3	5402	1.4	−13
Switzerland	4009	1.1	4211	1.1	−5
Italy	3494	<1.0	4486	1.2	−22
N. Zealand	3062	0.9	3385	0.9	−10
Saudi Arabia	2902	0.8	3071	0.8	−6
Brazil	2464	0.7	2458	0.6	0
Scandinavia	2378	0.7	2992	0.9	−21
Israel	2333	0.6	2654	0.7	−12
Mexico	2332	0.6	2488	0.7	−6
30.-South Africa	2296	0.6	2537	0.7	−10
World total (190 states)	359,180	100	381,810	100	−6

The situation is that a majority of States have delivered Action Plans in time. The 1% de minimis clause was applied with respect to the 2009 situation, with the purpose of putting more pressure on countries with a powerful aviation sector. However, as ICAO has no authority on domestic traffic, some countries with sizeable air transport, like Italy, Brazil, South Africa, or Mexico, are not compelled to present Action Plans, although some of them have presented them on a voluntary basis. Uneven levels of traffic growth may change the distribution in future years.

An Action Plan is a document that identifies the steps that States have to give in a given period of time to reach the desired goals. ICAO Resolution A37-19 encourages States to present their Action Plans to ICAO with a description of their respective policies and measures, as well as their annual reports on CO_2 emissions from international aviation. Resolution A38-18 indicates that the States and the relevant organizations shall work through ICAO to reach a 2% annual average improvement in fuel efficiency until 2020 and 2% from 2021 to 2050. In the long term, the global objective is to keep net CO_2 emissions from international aviation at the same level as in 2020. ICAO encourages States to share information and cooperate with other States, and also to make public their respective Action Plans, considering the sensitive information they may contain. Published Action Plans are available in ICAO web page (http://www.icao.int/environmental-protection/Pages/action-plan.aspx). ICAO provides guidance material for the preparation of the Action Plans, and also some tools like the ICAO Fuel Savings Estimation Tool (IFSET).

9.5 Alternative fuels

Alternative fuels represent a substitutive product to fossil kerosene, reducing the dependency and decreasing also the CO_2 emissions. The ideal option would be a drop-in fuel that can be blended with kerosene, and consequently not requiring modifications in engines, fuel system, or logistic fuel distribution system.

The following is a broad classification of alternative fuels:

- Fossil fuels
  - CTL (Carbon to Liquid, Fischer-Tropsch)
  - GTL (Gas to Liquid, Fischer-Tropsch)
  - Hydrogen (extracted from hydrocarbons
- Renewable fuels
  - Ethanol (sugar fermentation)
  - Biokerosene (vegetal oils, algae)
  - BTL (Bio to Liquid, Fischer -Tropsch)
  - Hydrogen (by hydrolysis)
- Nonliquid fuels
  - Fuel cells (chemical reaction)
  - Solar energy

The special interest for the environmental impact of air transport is biofuels. In order to consider a certain fuel as a biofuel, it must comply with the following points:

- Coming from a living organism: biomass is the origin of the fuel (fossil fuels—oil and coal—are not biofuels).
- Renewable: it is the capability to renew thanks to biological processes (fossil fuels are not renewable)
- Sustainable: applied mainly to the whole set of processes: manufacturing, extraction, distribution, etc.
- Biodegradable: the capability of Nature to process and assimilate in case of accident.
- Scalable: possibility to be produced at an industrial scale.
- Carbon neutral: the CO_2 emitted in the combustion or manufacturing has previously been compensated. In the case of a fossil fuel, oil burning liberates the CO_2, which is trapped underground. On the other hand, in a biofuel, plants reabsorb during their growth the CO_2 emissions.

Different than in the case of fossil fuels, the CO_2 amount absorbed by the plants, during their growing phase, compensates partially the emissions produced by the biokerosene combustion. Considering the total energy consumption in the process (crops, harvesting, distillation, distribution), the total net balance is 50%–70% better than the CO_2 emissions of the fossil fuel. Any meaningful comparison should compute the total emissions along the life cycle of the product, including manufacturing, distillation, distribution, and consumption.

Production is concentrated in "drop-in" fuel, directly mixable with kerosene and with similar characteristics. There have been a great number of tests in commercial flights by many airlines, using different kinds of biofuel and proportion of mixes (up to 50%, the maximum allowed by present engines certification). There is a biokerosene specification, prepared by ASTM International (American Society of Testing and Materials), approved in 2011, and a similar one in the United Kingdom.

Probably the most popular of the experiences using biofuels is the Lufthansa Pure sky Project. From 15th July to 15th December 2011, Lufthansa set the World's first regular scheduled commercial flights using biofuels. The experience consisted of four daily round trips between Hamburg and Frankfurt, with a total of 1200 flights during the 6 months of the project. The aircraft used in this experience was an Airbus A321 with IAE V2500 engines. Aircraft fueling was performed in Hamburg only, within Minimum Ground Time. The fuel consisted of a 50% blend of HEFA kerosene supplied by Neste Oil in one engine. The biofuel came from a mix of feedstocks: jatropha and tallow oil. Being part of the BurnFair R&T Project from the European Commission (joint research activity of 12 universities and industry partners under Lufthansa leadership), the purpose of the experience was to research on biofuel usage and engine performance.

The initial economic evaluations are very negative. High quantities of production are needed in order to reach economies of scale. There is a lack of long time operation data and figures on the engine reliability and maintainability effects in the long term. Modern engines are certified for taking up to 50% kerosene mixed with drop-in biofuel, but manufacturers give no reliability guarantees in that condition. There is a need of high initial investment levels in crops and distillation, including special technology for cultivation and harvesting.

Biofuels show however some additional advantages:

- Environmental: CO_2 emissions reduction without competing with the production of food for people or animals. The specification may be cleaner than fossil kerosene, taking out sulfur and aromatics.
- Strategic: diversifying fuel production sources from chemical and geographic points of view, reducing the oil producers dependency.
- Economic: stabilizing fuel price, reducing the typical oscillations of the oil market.

The aim to reduce CO_2 emissions 50% with respect to 2020 between 2020 and 2050 relies largely on the development at large scale of biofuels, as seen in Fig. 9.6. ICAO unveiled in October 2017 at its second Conference on Aviation and Alternative Fuels (CAAF/2) in Mexico City its long-term vision for the development, production and supply of sustainable aviation fuels (SAF) through to 2050, including proposals for short-, mid-, and long-term goals that ensure a 2% share of SAF in international aviation fuel demand by 2025, rising to 32% in 2040 and 50% by 2050. Due to the opposition from some States, the industry, and NGOs to back these specific targets, the Conference finally called for a "significant percentage" of SAF by 2050.

The development of biofuel technology for commercial aviation use made extraordinary advances in the last 6 years. In February 23, 2008, a Virgin Atlantic B747/400 with General Electric CF6-80C2 engines became the first commercial aircraft in making a flight with a mix of 80% conventional kerosene and 20% first generation biofuel, produced from coconut oil and seeds of babassu palm tree. Since then, >40,000 flights have been conducted using sustainable fuels and a number of airports are now offering such fuels to airlines interested in purchasing them. The results from the technical point of view have been satisfactory.

Commercial aviation kerosene has a very demanding specification with tight limits for the elements contained in the fuel and strict levels of energy content, density, viscosity, flash, and freeze point. From the beginning of the testing, the technical option was to produce biofuel with a high capacity of blending without changing the properties of the fluid, what in aeronautical terms is known as *drop-in* fuel. In this way, modification was needed neither in aircraft engine and systems nor in the logistic supply system, delivery, and emergency equipment at the airports.

On July 1, 2011, the American Society for Testing and Materials (ASTM International), the most recognized body on international standards, approved a biofuel specification based on Hydroprocessed Esters and Fatty Acids (HEFA), with the number ASTM D7566. An overwhelming majority of today's in-service turbine engines can operate with a mix of up to 50% of this specification biofuel in the standard kerosene.

Having overcome the technical feasibility problems, the EU accepted the challenge of promoting biofuels on air transport energy consumption. In June 2011, the European Commission, Airbus, and high-level representatives of the Aviation and Biofuel producers industries launched the private-public program European Advanced Biofuels Flightpath (EABF) with the purpose of achieve 2 million tons of sustainable biofuels used in the EU civil aviation sector by the year 2020. This action requires the work of every stakeholder to promote production, distribution, storage, and use of sustainably produced and technically certified biofuel. For this, it is necessary to join forces in establishing appropriate and effective financial mechanisms to support the construction of industrial "first of a kind" advanced biofuel production plants.

The project actions focus on the following issues:

- Facilitate the development of standards for drop-in biofuels and their certification for use in commercial aircraft.
- Work together with the full supply chain to further develop worldwide accepted sustainability certification frameworks.
- Agree on biofuel take-off arrangements over a defined period of time and at a reasonable cost.
- Promote appropriate public and private actions to ensure the market uptake of paraffinic biofuels by the aviation sector.
- Establish financing structures to facilitate the realization of 2G biofuel projects.
- Accelerate targeted research and innovation for advanced biofuel technologies, and especially algae.
- Take concrete actions to inform the European citizen of the benefits of replacing kerosene by certified sustainable biofuels.

The working methods and governance will be the ones established in the Strategic Energy Technology Plan (SET-Plan) of the EU, approved on February 4, 2011. This initiative falls under the European Industrial Bioenergy Initiative (EIBI) included in SET-Plan.

EABF proposal selects three biofuel processes as main candidates for mixing with aviation kerosene:

- Synthetic Fischer-Tropsch (FT) kerosene, produced via lignocellulosic biomass gasification followed by gas cleaning and synthesis over appropriate catalysts, approved by ASTM D7566 for a 50% blend.
- Hydrogenated Vegetable Oils (HVO), based on triglycerides and fatty acids, which can originate from plats oil, algae, and microbial oil. In absence or technical restraints, market forces and legislation are the main forces for oil and fat selection.
- Hydrogenated Pyrolysis Oils (HPO), based on pyrolysis oils from lignocellulosic biomass. Pyrolysis oils can be hydrotreated either in dedicated facilities or coprocessed with petroleum oils in refineries.

A general discussion of the state of the art in different processes can be seen in the presentations of a Workshop held in Milan on June 18, 2012, under the title "Upstream R&D and Innovation for biofuels in aviation." There were presentations on Fischer-Tropsch fuel chain, alcohol to biojet fuel, pyrolysis and hydrothermal, microbial oils, HVO, and algae. A typical scheme of an HVO process can be seen in Fig. 9.7.

The market for transport biofuel can be split into two groups:

- First-generation biofuels are commercially available, using dedicated feedstocks, such as sugar beet, oilseeds, and starch crops. Sugars in these crops are fermented to produce ethanol, while oil crops provide oil that is trans-esterified to form fatty acid methyl ester. The resulting ethanol and biodiesel are then generally mixed with fossil-based liquid fuels.
- Second-generation biofuels are generally not yet commercially viable but are expected to play an increasing role in the coming decades. They use mainly lignocellulosic feed stocks, e.g., short-rotation coppice, perennial grasses, forest residues, and straw, but also some oil seeds not used for food, like jatropha or camelina. They can be treated by thermochemical

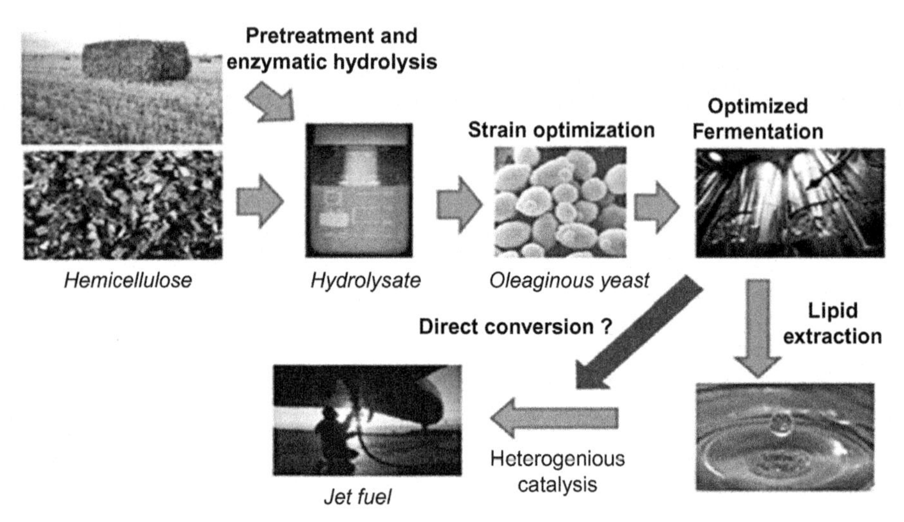

Fig. 9.7 The yeast oil to biojet chain.

or biochemical conversion. Currently, there are no commercial biomass-to-liquid plants but several precommercial plants exist in Finland, Germany, Japan, and the United States. Biochemical conversion involves pretreatment of cellulosic biomass and enzymatically enhanced hydrolysis and subsequent fermentation to convert hemicellulose and sugar into ethanol. There are demonstration plants in the EU (Denmark, Spain, and Sweden) and Canada. Other countries such as Brazil, China, Germany, Japan, and the United States are also developing such second-generation ethanol technologies.

Some of other promising possibilities like algal oils or direct conversion of sugars to hydrocarbon fuels are either very high-cost processes or are still in the early research phase, being unlikely of playing any significant role by 2020.

The technical certification of aviation biofuels is co-ordinated among the US based ASTM International and the UK DEF STAN organization for Europe, producing identical specifications. FT fuels have been approved in ASTM D7566 Annex 1 and HVO fuels approval is in Annex 2, while HPO are still being tested. Similar provisions are contained in DEF STAN 91-91 Annex D. During the certification process, intensive tests have been conducted both by airframe and engine manufacturers.

Since through the ASTM International approval process, certain types of biokerosene and blends with conventional kerosene are recognized as meeting the conventional jet fuel ASTM 1655 specification, the existing infrastructure (most importantly pipelines, but also trucks and ships) can be used both for transport to and for fueling at the airport. Thus, the supply and logistics of certified drop-in biofuels, whatever the blend, presents no problem to the distributors in this respect.

Each one of the mentioned three processes has a different cost structure. HVO requires a modest upfront capital investment but production cost is highly dependent on vegetable oil feedstock prices, which can make up to 60%–75% of the final product price. Lignocellulosic FT and HPO biofuels have relatively low raw material cost, but more expensive processes, requiring high initial investments that makes a very expensive product at the beginning of the manufacturing operation. Hypothetically, the prices will get down as the scale of the production increases.

The price setting and the associated support policies need to take into account all the three main benefits of aviation biofuels:

(a) Reduction of external emissions of GHGs in order to reduce the climate change effect of aviation.
(b) Security of supply in order to reduce and differentiate EU energy supply channels from abroad.
(c) Improvement in rural income and conditions in order to reduce existing gaps within the EU and to create new outlets for farm production.

All these benefits should be achieved complying with the sustainability criteria and without altering the balance of the EU internal market.

There is little doubt that present technology can create a fuel that replicates standard aviation kerosene from sustainable alternative raw materials, but doing so in a cost-effective way and at the scale of the industry needs remains a formidable challenge. According to recent EU estimations, aviation biofuel may cost above 2000 euro per tonne, against 700 euro per tonne of fossil kerosene at today's prices. Obviously the European airline industry cannot afford trebling the price of an element representing between 30 and 35% of its total costs without losing competitiveness to other world areas airlines.

This relative position may change as the oil becomes scarce and its price goes up, but some experts suggest that feedstock prices are going up as well in not a very different way than oil, as seen in Fig. 9.8.

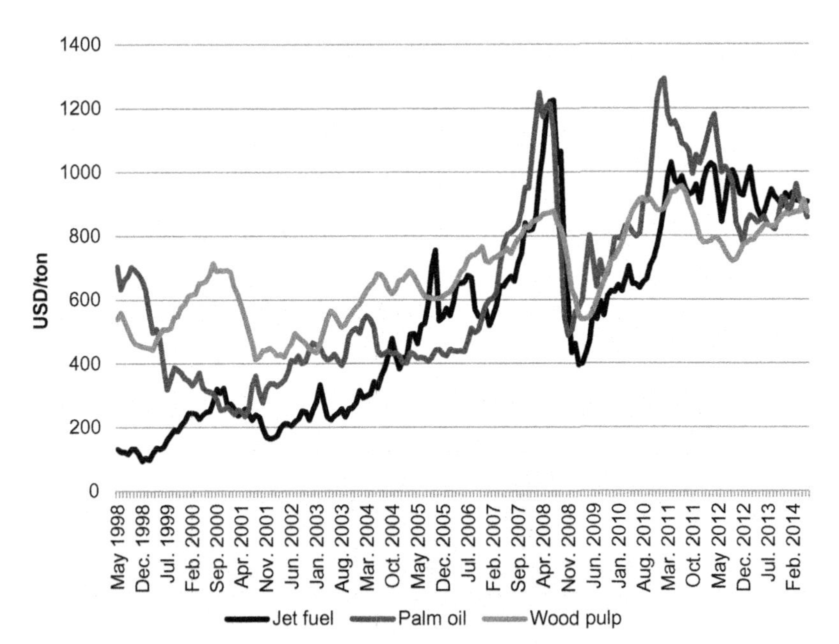

Fig. 9.8 Cost of Jet kerosene versus selected biofuel raw materials 1999–2014 (*Adapted and used with data from https://www.indexmundi.com/commodities/*).

The evaluation of the three external effects mentioned is required. The CO_2 reduction may be evaluated at first sight considering the European Emissions Trading System (ETS) that gives biofuels a zero emission factor. Then biofuel consumption is practically exempted of the emissions market. Burning a tonne of kerosene emits 3.16 t of CO_2 that if replaced by biofuel will not have to buy any emission allowance. Unfortunately, the low price of allowances would translate this in 25 euro savings at a price of 8 euro per emitted ton. Present predictions for future CO_2 allowances price do not go much higher unless world States having ratified the International Framework Convention on Climate Change (IFCCC) agree on more strict rules in the future Conference of the Parties (COP) yearly meetings.

Security of supply is more difficult to evaluate. Additional supplying sources may limit the price increase of the oil as more providers would enter into the market and get up the competition level, but aviation kerosene is taking around 7% of the oil production and the European part will be close to 2% of the total oil extraction, very low fraction to have an immediate price repercussion.

The same reasoning is applicable to the additional jobs and wealth created by local production. The total EU renewable energy target for transport increases crop area globally by 5.2 million hectare, equivalent of 0.7% of the cultivated European surface. The economic effect of this action will not be negligible, but it remains to be seen where to find fertile idle land that, by the way, is not a renewable resource. Expected biomass sources for the 2006–20 period can be seen in Fig. 9.9 from that reference.

The present projections of the National Renewable Energy Action Plans (NREAPS), established by the Directive 2009/28/EC, forecast a 50% of the agricultural bioenergy in 2020 coming from farming (the so-called energy crops) and the other 50% being obtained from agricultural waste and residues. This second part would grow much faster as indicated in Table 9.6 from the same reference.

As a conclusion, the successful introduction of aviation biofuels will depend on the selection of the right support schemes at European level in order to promote the most

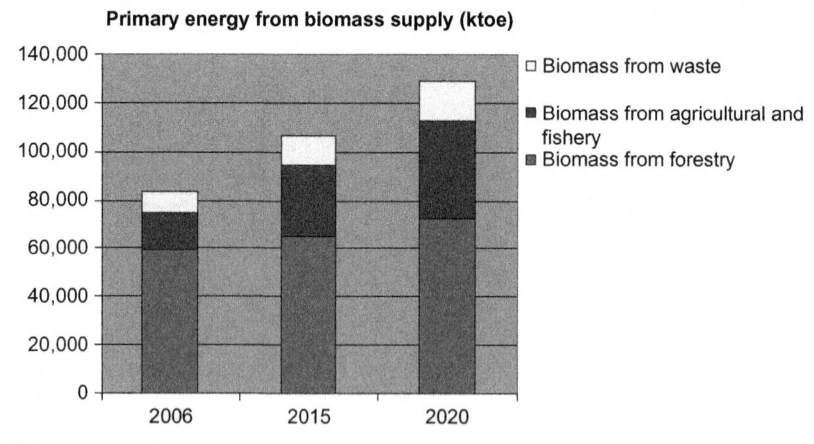

Fig. 9.9 Expected biomass sources from the NREAPS in thousands of tonnes of oil equivalent.

Table 9.6 EU biomass production in millions of tonnes of oil equivalent

	2005	2030	2050
Crops	5	53	134
(of which 2nd-generation crops)	*0*	*40*	*127*
Agricultural residues	17	32	49
Forestry	40	51	59
Waste	25	63	87
Import	2	12	26
TOTAL	90	212	356

efficient type of production in both environmental and economic terms. This does not disregard the possibility of reaching a biofuel production at great scale that makes it more competitive with fossil fuels but it might take a time period out of the 2020–30 scenarios presently contemplated and there is no indication that support schemes may become unneeded in the foreseeable future.

It is widely accepted that the introduction of biofuels cannot be done using current market mechanisms because their high production and distribution cost, at least during the initial stage, limit their competitiveness with fossil fuels. Then, all States interested in promoting the use of biofuels are going to use special incentives to gain public acceptance of these new products.

There are several feasible options for incentive policies, varying on the type of economic mechanism applied, subsidies or quotas, and in the phase of the production-consumption cycle where it is applied to, farming, recollection, manufacturing, distribution, or consumption. Each State or even part of it, like the federal States of USA and Canada, may select the combination of measures that fit the best to its particular environment and produce less interference with other related activities, like food agriculture, rural development, or other energy production.

In the United States, there were some legislative actions to launch alternative energy production, like the 2005 Energy Policy Act, the 2007 Energy Independence and Security Act, and the 2008 Farm Bill. Biofuels, in this case, transportation fuels produced from plants and other organic materials, are of particular interest. Ethanol and biodiesel, the two most widely used biofuels, received significant federal support in form of tax incentives, loan and grant programs, and regulatory programs Table 9.7 listed 22 recent Federal programs that provide direct or indirect incentives for biofuels, grouped by administering agency. Some of them are long term, some have a fix deadline, and others need Senate or Congress approval for continuity.

As seen, the support schemes are being administrated by six different agencies and departments without any apparent coordination. Many of them have been originated by three major Federal legislative actions previously mentioned, the 2005 and 2007 energy laws and a farm bill in 2008. A great number of them are temporary rules terminating at the end of 2011 or 2012 and will need Congress approval to be extended.

Table 9.7 **Summary of US Federal incentives promoting biofuels (Congressional Research Service)**

Administering agency	Program	Description
Environmental Protection Agency (EPA)	Renewable fuel standard	Mandated use of renewable fuel in gasoline: 4 billion gallons (2006), 36 billion gallons (2022)
Internal Revenue Service (IRS)	Volumetric ethanol excise tax credit	Gasoline suppliers who blend ethanol with gasoline are eligible for a tax credit of 0.45 USD per gallon of ethanol
	Small ethanol producer credit	An ethanol producer with <60 million gallons per year in production capacity may claim a credit of 0.10 USD per gallon on the first 15 million gallons produced in a year
	Biodiesel tax credit	Producers of biodiesel or diesel/biodiesel blends may claim a tax credit of 1 USD per gallon of biodiesel
	Small agribiodiesel producer credit	An agribiodiesel (produced from virgin agricultural products) producer with <60 million gallons per year in production capacity may claim a credit of 0.10 USD per gallon on the first 15 million gallons produced in a year
	Renewable diesel tax credit	Producers of renewable diesel may claim a tax credit of 1 USD per gallon of renewable diesel
	Credit for production of cellulosic biofuel	Producers of cellulosic biofuel may claim a tax credit of 1.01 USD per gallon, under some specific conditions
	Special depreciation allowance for cellulosic biofuel plant property	Plants producing cellulosic biofuels may take a 50% depreciation allowance in the first year of operation, subject to certain restrictions
	Alternative fueling station credit	A credit up to 30,000 USD is available for the installation of alternative fuel infrastructure, including E85 (85% ethanol and 15% gasoline) pumps
Department of Agriculture (DOA)	Biorefinery assistance	Loan guarantees and grants for the construction and retrofitting of biorefineries to produce advanced biofuels
	Repowering assistance	Grants to biorefineries that use renewable biomass to reduce or eliminate fossil fuel use

Table 9.7 Summary of US Federal incentives promoting biofuels (Congressional Research Service)—Cont'd

Administering agency	Program	Description
	Bioenergy program for advanced biofuels	Provides payments to producers to support and expand production of advanced biofuels
	Feedstock flexibility program for producers of biofuels	Authorizes the use of CCC funds to purchase surplus sugar, to be resold as a biomass feedstock to produce bioenergy
	Biomass crop assistance program (BCAP)	Provides financial assistance for biomass crop establishment costs and annual payments for biomass production; also provides payments to assist with costs for biomass collection, harvest, storage, and transportation
	Rural energy for America program (REAP)	Loan guarantees and grants for a wide range of rural energy projects, including biofuels
	Biomass Research and Development	Grants for biomass research, development, and demonstration programs
Department of Energy (DOE)	Biorefinery project grants	Funds cooperative R&D on biomass for fuels, power, chemicals, and other products
	Loan guarantees for ethanol and commercial byproducts from various feedstocks	Several programs of loan guarantees to construct facilities that produce ethanol and other commercial products from cellulosic material, municipal solid waste, and/or sugarcane
	DOE Loan guarantee program	Loan guarantees for energy projects that reduce air pollutant and greenhouse gas emissions, including biofuels projects
	Cellulosic ethanol reserve auction	Authorizes DOE to provide per-gallon payments to cellulosic biofuels producers
U.S. Customs and Border Protection	Import Duty for fuel ethanol	All imported ethanol is subject to a 2.5% ad valorem tariff; it is also subject to a most-favored-nation added duty of 0.54 USD per gallon
Department of Transportation (DOT)	Flexible fuel vehicle production incentive	Automakers subject to Corporate Average Fuel Economy (CAFE) standards may accrue credits under that program for the production and sale of alternative fuel vehicles, including ethanol/gasoline flexible fuel vehicles (FFVs)

The schemes can be broadly divided into six types:

- A mandate for the use of fixed amounts of renewable fuel in gasoline
- Tax credits for gasoline suppliers and biofuel producers,
- Financial help for the opening and expansion of supply and production facilities, either new or refurbished
- Financial help for Research and Development
- Reduction of import duties for some kinds of biofuels
- Incentives to manufacturers for production and sale of alternative fuel vehicles

but can be resumed in two groups: mandate for blending and tax measures.

The distribution of incentives goes along the whole biofuel chain, from the agriculture to the vehicle, passing through the production plants, the refineries and the distribution, with selective inclusion of some products (fuel ethanol) in an import duty rebate. All these policies have a cost for the tax payer, diminishing the public treasure revenues. No information has been published on that amount.

The second-generation biofuels for transport are included in the Renewable Fuel Standard (RFS) law. President Obama supported the preservation of the RFS, as a part of an "all of the above energy strategy," based on a public-private finance scheme. However, there is fear that affordable private capital will not be available to support any major capacity building for advanced biofuels, putting the RFS itself, with its steep annual volumetric increases, in considerable jeopardy. The resulting lack of capacity and rewriting of mandates to support lower levels of capacity building may result in market uncertainty and further reduce investor interest in the sector.

A large variety of biofuel support policies are in place in EU Member States, ranging from command and control instruments such as standards and quotas, over economic and fiscal measures, such as tax exemptions, to information diffusion, and addressing different stages of the biofuel chain.

As first-generation biofuels are a mature product, the policy focus today lies on facilitating their market entry rather than R&D support. As the production cost of biofuels is well above those of fossil fuels, incentive policies are needed to create the biofuels market. This is done, as in the United States, by two types of instruments: subsidization or prescription of a mandatory production. In the first case, a tax reduction scheme brings the price of biofuel down to the price of the equivalent fossil fuel, reducing the State income. In the second case, fuel providers are obliged to blend a certain percentage or gross amount of biofuel in their product. In this way, fuel providers and the transport users carry the additional cost and a certain reduction of demand is caused. Similar schemes may be applied to other elements, like fiscal support to the sale or utilization of dedicated vehicles, either in their sale price or in the owner taxes or the circulation conditions, e.g., allowing biofueled vehicles to circulate by restricted lanes or to have access to the big cities downtown.

Unlike the United States, the EU has tried to get a higher level of coordination in the State aid policies, defining global targets and standardizing legal and economic procedures, but leaving Member States some open space to determine the level of each policy application and the mix of support schemes more adequate to their particular characteristics. In any case, the used tools are very similar to those used in other countries, divided into fiscal measures and blending obligations.

The legal framework for the fiscal measures is the EU Energy Taxation Directive 2003/96/EC that allows exempting all types of biofuels from taxes if:

- The tax exemption or reduction does not exceed the amount of taxation payable on the volume of renewables used.
- Changes in the feedstocks prices are accounted for in order to avoid overcompensation.
- The exemption or reduction authorized may not be applied for a period of more than six consecutive years, renewable.

In April 2014 the European Commission introduced new guidelines on state aid for renewable energy, including biofuels, with the purpose of providing a framework for designing more efficient public support measures that reflect market conditions. According to the Commission Vice President in charge of competition policy, Joaquín Almunia, Europe should meet its ambitious energy and climate targets at the least possible cost for taxpayers and without undue distortions of competition in the Single Market.

Three months before, on January 23, 2014, the Commission had published a communication titled: *A policy framework for climate and energy in the period from 2020 to 2030.* In relation to biofuels, the communication confirmed the relevancy of the 2020 targets (10% of renewable fuels for the transport sector) and acknowledges the achieved progress, with a 4.7% rate in 2010 compared with a 1.2% in 2005. Based on the past experience, the Communication concluded that first-generation biofuels have a limited role in decarbonizing the transport sector and a range of alternative renewable fuels and a mix of targeted policy measures building on the Transport White Paper are thus needed to address the challenges of the transport sector in a 2030 perspective and beyond. No specific 2030 target was announced.

The April 2014 guidelines on State aid made reference to biofuels for trying to progressively reduce support of first-generation biofuels (called food-based biofuels), without causing a sudden disruption on the industry. It considered that investment aid in new and existing capacity for food-based biofuels is not justified, in view of the existing overcapacity in this market, but this aid can be justified for increasing capacity in advanced biofuels or in the conversion of the existing facilities in these ones. Thus, investment aid to support fuel-based biofuels is stopped with the entry into force of these guidelines.

Operating aid for first-generation biofuels will cease at the end of 2020. This aid can only be granted to the production of plants that started operation before 31 December 2013 until the plant is fully depreciated but in any event no later than 2020.

With respect to biofuels subject to supply or blending obligations, aid can be provided if a Member State can demonstrate that the aid is limited to sustainable biofuels that are too expensive to come on the market with a supply or blending obligation only. The concept of "sustainable biofuel" refers to those fulfilling the sustainability criteria set out in Article 17 of the Directive 2009/28/EC.

During the second half of 2013, both the European Council and the European Parliament were extensively discussing the possibility of putting a cap on the first-generation biofuels utilization and how to include the ILUC factor in the accounting of environmental impact of biofuels. No agreement was reached, but the issue is expected to appear again after the recent European Elections, with the goal of achieving a final decision at the end of 2014.

With respect to the blending obligations, each country may apply its selected amount, based on percentage of energy or volume, making it indicative or obligatory or minimum level. For example, Estonia, Greece, and Lithuania have only indicative blending amounts and Denmark adopted a first-ever obligatory quota in 2011. Those blending levels may be different for biodiesel and bioethanol, and have been slowly increasing during the last years. Table 9.8 shows the status of blending obligations in the EU and Norway (associated State in bioenergy policies) in the year 2011. Sales of biofuels reached 13.9 million tonnes; a 77% of biodiesel, 21.5% of bioethanol, and 1.5% of vegetable oil. As indicated before, present mandate is 5.75% and the final target is 10% of renewable content in 2020.

USA and Europe are not the only countries with mandates. A total of 13 American States, including Argentina, Brazil, Canada, Chile, Costa Rica, Ecuador, Jamaica,

Table 9.8 Blending obligations in the EU, measured in % of energy

Member state	Total %	Biodiesel %	Bioethanol
Belgium	4	–	–
Bulgaria	5[a]	4[a]	2[a]
Czech Republic	–	6[a]	4.1[a]
Denmark	3.5	–	–
Germany	6.25	4.4 (min)	2.8 (min)
Estonia	5.75 (indicative)	–	–
Ireland	4[a]	–	–
Greece	5.75 (indicative)	–	–
Spain	7	6 (min)	3.9(min)
France	7[a]	7[a]	7[a]
Italy	4	–	–
Cyprus	2.5	–	–
Latvia	5.75[a]	–	–
Lithuania	5.75 (indicative)	–	–
Luxembourg	–	–	–
Hungary	5.75	4.4[a]	4.4[a]
Malta	–	–	–
Netherlands	4.25[a]	3.5	3.5
Austria	5.75	6.3 (min)	3.4 (min)
Poland	6.2	–	–
Portugal	5	6.75[a]	–
Romania	5.75	5[a]	5[a]
Slovenia	5.5	–	–
Slovak Republic	5.75	5.2[a] (min)	3.1[a] (min)
Finland	6	–	–
Sweden	5.75 (indicative)	–	–
United Kingdom	4[a]	–	–
Norway (non-EU)	5[a]		

[a] Indicates % by volume.

Mexico, Panama, Paraguay, Peru, Uruguay, and the United States itself, have different blending obligations, putting more emphasis on ethanol than in biodiesel. Some federal countries, like Canada, use different figures in each one of its States.

In the Asia-Pacific region, other 12 States have implemented this kind of policies, namely, Australia, China (in nine provinces only), Fiji, India, Indonesia, Malaysia, New Zealand, The Philippines, South Korea, Taiwan, Thailand, and Vietnam. Ten African States, Angola, Ethiopia, Kenya, Malawi, Mozambique, Nigeria, South Africa, Sudan, Zambia, and Zimbabwe, have ethanol blending obligations, but South Africa has a biodiesel mandate as well. A total of 62 of the world larger economies, with the notable absences of Japan, Russia, and the Arabian Gulf States, are in the list. The highest consumers Brazil, China, EU, and United States are targeting average mixes of 10%–15% in the 2020–25 period.

In the proceedings of the Workshop on March 20, 2012 in Brussels, there are some examples of financial aid possibilities to build up second-generation biofuels production capacity. There is a general consensus on the need of public financial aid for Research and Development and on the convenience of public-private consortia for the initial phases of the production. The banking sector offers a very uneven perspective. Some public banking institutions, like the European Investment Bank (EIB), are lending considerable sums to renewable energy projects. EIB itself lent 4500 million euros in 2011, although the largest part went to well-established sectors, like wind energy.

The private banking is a bit hesitant to finance biofuels because of their financing experiences with the first-generation biofuel projects, suffering of current overcapacity and dependency on frequently changing policies, sometimes in different countries around the world, but often within the same State, according to the fluctuations of the economy.

The key success factors that any bioenergy project should bring to the table, when requesting for bank financing are:

- Logistic and location.
- Price risk management.
- Feedstock supply with easy and assured access.
- Assured take-off contracts.
- Convincing capacity utilization expectations (equal to or higher than 75%).
- Bring in experience management.
- Comply with sustainability criteria.

Different support schemes have different characteristics, adjusting better or worse to the local market conditions. It is risky to pick up one as the optimum for worldwide application. With the experience gained during the last years in the United States and EU markets, it is possible to reach some initial approach to the virtues of each mechanism, by reviewing one by one their main features.

Indicative/Mandatory targets: Indicative targets give more flexibility for taking into account the particular conditions of different countries and for adapting to the changing results of yearly crops. Mandatory largely determines the target credibility, meaning more efforts by the governments to achieve the targets. This in turn means that the markets have greater certainty for planning and undertaking investments.

Volumetric/Percentage targets: volumetric mandates reduce uncertainty faced by biofuels producers and farmers. Percentage targets address the commitment of the sector in meeting the overall greenhouse emissions of Kyoto Protocol limits. The effectiveness of either type of mandate is conditional on its interaction with any other policy or technical restriction that is in place and that impact biofuel use. For instance, older road vehicles cannot use higher blends and limit the applicable percentage of biofuel.

Incentives for second-generation biofuels/Quantitative distribution of mandate by type: the first approach, used by the EU, is more flexible in that the volume of second-generation biofuels is related to their competitiveness with respect to other biofuels. The US approach provides more certainty by setting a mandate volume, but this mandate can be waived depending on technology available and this undermines its theoretical advantages.

Supranational/Country-level strategies: The current decentralized approach of the EU has the advantage of allowing countries to find the most effective means to achieve the targets, which may differ country by country. In the USA, it is the Renewable Identification Number (RIN, a 38-character numeric code that goes with each biofuel package, identifying its vintage, volume, and fuel classification) market that allows production and distribution to move to the most cost-effective facilities and regions.

Subsidies/Mandates: The subsidy system has been largely implemented in Europe to the first-generation biofuels, causing significant revenue losses for the governments. However, tax exemptions have the ability to steer the market by applying different reduction rates to various types of biofuels. With the mandatory obligation to blend, fuel suppliers are obliged to achieve a certain share of biofuels in their total fuel sales. This instrument does not cause any revenue losses for the government since the fuel supplier and final consumers are carrying the financial burden of this measure. The higher prices reduce transport fuel demand compared to tax-exemption scheme. There is currently a hectic controversy on whether a combination of obligation to blend and simultaneous tax relief may be the most effective instrument, as EU maintains. Other researchers suggest that a biofuel mandate is fully equivalent to a combination of fuel taxes and biofuel subsidies that is revenue neutral. According to this approach, the mandate alone would be sufficient and any further subsidies would be income transfers from taxpayers to customers.

With increasing demand and decreasing marginal fuel efficiency improvements, the contribution of aviation to climate change relative to other sectors is projected to increase in the future, but it is difficult to predict how much. Some analysis opens the incertitude margin for the EU ranging from 10% to 50% of all GHG emissions by 2050.

According to the IEA, biofuels could meet 27% of the transport fuel demand by 2050, with a key role played by advanced biofuels, which need to be rapidly deployed and scaled up. This task would need great and fast progress in three different areas:

- Ensure a stable supply of sustainable feedstocks
- Support demand by establishing a specific mandate for advanced biofuels by 2020 and beyond
- Establish mechanisms to finance first of their kind advanced biorefineries

In favor of this approach's success in Europe, there is a stepping up of the incentives for second-generation biofuels Research& Development and Innovation with funds provided by the Seventh Framework Programme (FP7) and Horizon 2020 programs, as the target of reducing GHG emissions of 80% by 2050 will require a substantial contribution from transport sector, implying an increase of biofuels to an 8 times' bigger amount than today.

One of the initiatives financed by FP7 funds is the Initiative Toward sustainable Kerosene for Aviation (ITAKA), a collaborative project aimed to produce sustainable renewable aviation fuel end to test its use in existing logistic systems and in normal flight operation in Europe. Led by a consortium of aerospace and fuel companies, with academic collaboration, ITAKA intends to make a contribution to the 2 million tonnes of aviation biofuel target by 2020, included in the Flightpath 2020 program.

Negative conditions are also important, starting for the strong controversy about the long-term use of first-generation biofuels, and the slow deployment of the second-generation biofuels production plants and a great undefinition of the type of feedstock and processes to be used. The significance of measures on the supply side may therefore increase as a tool to steer a growing biofuel market into the desired direction, yet an additional cost compared to a least-cost approach. A crop specific feedstock support subsidy like the energy crop scheme may help to direct the crop mix into an environmental and landscape-beneficial pathway.

A new and very interesting point is the standardization of the Indirect Land Use Change (ILUC) evaluation techniques that may allow to establish target in terms of climate change effects, like equivalent CO_2 emissions, instead of mandates based on volume, weight, or energy. At this moment, there is no agreement on how to calculate ILUC effect with good level of accuracy and some ecologist groups are suggesting that certain types of feedstocks produce biofuels with very low life-cycle CO_2 savings. This will make it undesirable from an environmental point of view, although it may be considered positive from an economical or supply security approach.

The air transport sector can take a good note of the lessons of other transport modes experience. It has some advantages from the technical point of view, because there is a great homogeneity in the vehicles and engines and in the specification of the used fuel. The selection of a drop-in policy makes very clear the feature of the biofuels to be used by this sector. This puts the so-called blend wall (the maximum percentage of blended biofuel that is acceptable by the operators) at 50%, a quantity much higher than in road traffic. At the same time, the specific features of aviation kerosene do not allow the use of most of the current first-generation biofuels and strongly reduce the candidate feedstock numbers to a few types (jatropha, camelina, urban waste, algae, and some others) of second or third generation.

The existence of EU Emissions Trading System that gives biofuels a zero emission factor may be taken as an incentive for biofuel use. However, the low present price of carbon allowance makes it not very significant. At the same time, the present air transport fuel distribution and logistics system makes it difficult to control, through the existing monitoring, reporting, and verification system, the credits that specific airlines should be given. The likely change of the present regional ETS to a worldwide system, perhaps based on an offsetting mechanism, can be influential in the re-evaluation of this incentive.

One additional problem may be the competition with other transport modes. If road transport has to achieve 10% of renewable energy by 2020, that may cause a shift of all available sustainable biomass to road transportation biofuels production, in particular if first-generation biofuel production is limited or eliminated.

Policy mechanisms for supporting alternative fuels in aviation should comply with the following principles:

- Parity with other economic sectors
- Maintain the competitiveness of this transportation mode
- Ensure high sustainability standards
- Temporary support mechanisms
- Focus on reducing investment risk

The application of blending obligations is more difficult in the air transport sector than in other activities, due to the international character of aviation. A mandate in some countries may imply fuel price differentiation and deviation of connecting traffic through airports not included in the mandate. Also the airlines themselves may have to pay different amounts for fuel depending on the practical implementation of the mandate (by airports, by regions, by countries, by continents). This increases the convenience of applying provisionally mandates by airlines until the biofuel market is stabilized, the use of tradable certificates, as the RINs in USA is other possibility to be considered.

Tax exemptions or subsidies seem to be easier to apply in the production and distribution phases than to the operator, leaving aside the already-commented ETS case. At the moment, the gap between the price of the fossil kerosene and biokerosene is so dramatic that it makes almost unthinkable to cover it with public money. However, tax exemptions or financial grants for R&D and start-ups plants seem an acceptable option during the first years of second-generation biofuels initial high-scale production.

9.6 Product life cycle

The modern concept of the environmental impact of air transport affects not only the aircraft operation, but the so-called Product Life Cycle, i.e., all the activities from aircraft manufacturing to their final disposal. Large aircraft manufacturers such as Airbus and Boeing, but also airlines and other agents in the industry, especially airports, are taking steps along this process of considering the environmental implications of the full product life cycle.

For instance, the Airbus environmental life-cycle approach is shown in Fig. 9.10 and consists basically of the following main lines of action:

- Investing in research to design cleaner and quieter aircraft
- Managing the supply chain for a shared vision of environmental responsibility
- Managing the impact of manufacturing on the environment thanks to cleaner technologies and processes
- Optimizing aircraft operations and maintenance for enhanced environmental performance
- Implementing new best practices to disassemble and recycle end-of-life aircraft

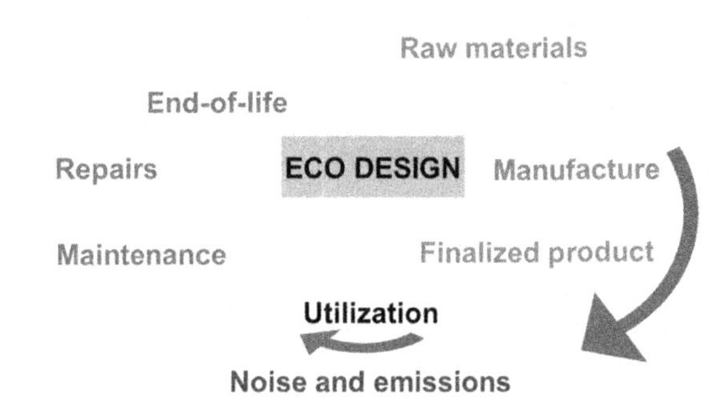

Fig. 9.10 Need to consider the total life cycle.

Another example related to Product life cycle is TARMAC (Tarbes Advanced Recycling & Maintenance Aircraft Company), a European platform with the participation of Airbus complying with applicable regulations related to environment, health & safety, and airworthiness using dedicated zones and infrastructures based upon business-related risk analyses. The Platform activities and services are parking & storage, light maintenance, part-out and storage of components, and smart and selective dismantling.

Boeing is also investing to improve the fuel efficiency and environmental performance of their products, services, and operations. The company takes into account environmental performance at every step of a product's lifecycle, from materials, design and manufacturing, through in-service use and end-of-service recycling and disposal, calling this strategy Design for Environment.

In 2006 Boeing and 10 other companies established the Aircraft Fleet Recycling Association (AFRA). There are now >40 member companies from 11 countries. AFRA is committed to continuously improving aircraft recycling methods' efficiency and environmental benefits. By working to efficiently process as many aircraft as possible, AFRA member recyclers make recycling more cost effective for aircraft owners (an aircraft manufacturer is not an aircraft recycler itself.) This will ultimately help ensure that aircraft recycling has an economically viable future in the marketplace. Collectively, AFRA member organizations have already recycled approximately 6000 commercial aircraft and approximately 1000 military aircraft (800 tactical) and also remarketed (returned to service) approximately 2000 airplanes.

AFRA's goal for its certified members is to recycle 90% of each aircraft by end of 2016. This includes safe and economical return of aircraft, engines, and parts to revenue service as well as return of reclaimed metallic and composite materials back into commercial and aircraft manufacturing.

An example of recycling is the Boeing agreement with InterfaceFlor to develop carpet tiles made from 100% recycled aerospace carpeting. Starting in March 2012, Southwest 737s began receiving their new planes off the Renton Assembly lies with the carpet tiles installed. Recycled carpet tiles will be soon an option available to customers on all Boeing airplanes. By using carpet tile instead of wall-to-wall carpeting, airlines replace only those sections that get stained and that maintenance can be completed overnight. The carpet tiles that are replaced can be recycled over and over again. In most airplanes, the carpeting is replaced 20 to 30 times over its lifetime. On a 777, that can add up to 18 tons of carpeting sent to landfills if the airplane is in service for 20 years. Given that aerospace carpeting is made from a synthetic material that is designed to be durable and meet aerospace fire-retardant properties, it doesn't break down readily in landfills.

9.7 Environmental management system

The purpose of Environmental management in a company is to make sure that the environmental impact of all the company activities is well known, measured, and controlled in order to prevent, limit, or reduce their impact.

The ultimate target of the company's environmental management is to increase its capability of affording the different demands in financial, economic, political, and social terms that society is going to pose.

The environmental management in companies has followed an evolution like the one depicted in Fig. 9.11.

The first phase can be considered of Resistance. Before 1980, companies did not recognize the need of taking into account environmental costs. Environmental protection was considered a labor health problem. Regulation was strictly normative, with exclusively technical tentative solutions. As a consequence, there were no environmental impact measurement and monitoring systems.

During the 1980s, the Compliance phase, regulations became more flexible, oriented toward practical results. Public Administrations created organizations specialized in environmental policies and strategies. Environmental impact measurement and monitoring systems were for the first time implemented, having some immediate results in the more direct effects. Also during this period the first company environmental reports were published.

During the 1990s, a "Win/Win" approach was introduced, with the assumption that "Green" initiatives may produce competitive advantages. Priority was given to the most profitable aspects of environmental policies: those with the best cost-benefit ratio. Companies invested in training and applied research and a great importance was given to the public opinion. The concept of Corporate Social Responsibility appears for the first time.

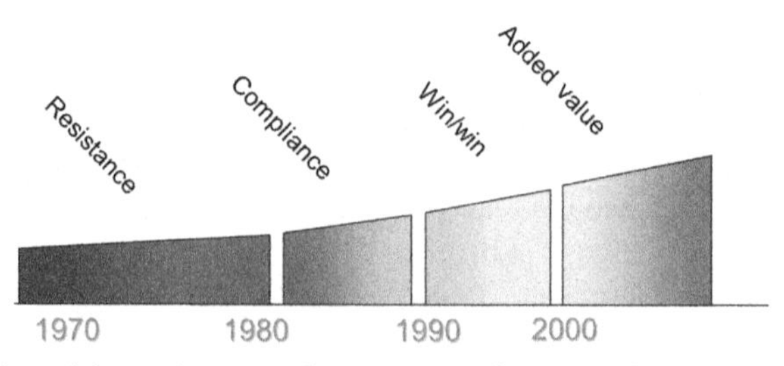

Fig. 9.11 Evolution of the environmental management in companies.

In the 21st century, environmental impact starts to be considered as a "horizontal" matter, affecting all the company activities. Legal regulatory compliance is given for granted as a basic issue in the business plans. Each Department may select its own environmental targets, based on cost-benefit analysis. The search of potential added value coming from environmental activities begins.

The concept of added value in the environmental field has different approaches:

- Economic approach: added value is created when negative environmental and social impacts of one activity are more than compensated by socioeconomic benefits
- Pragmatic approach: added value is created if environmental policies achieve when this factor does not limit the operability or the productive system development
- Environmental approach: added value is created only if the environment (natural and human) is able and willing to accept a particular activity

There are many reasons to support environmental management in a company, for instance:

- Satisfy the customers, employees, and shareholders' wishes
- Reduce direct operating costs
- Gain competitive advantages
- Obtain authorizations for business growth
- Increase public acceptance
- Fight against negative regulatory proposals
- Prevent operating restrictions
- Limit the costs related with regulatory compliance
- Build a positive Corporate Image
- Contribute to Sustainable Development

The inputs to develop an Environmental Management system in a company are shown in Fig. 9.12. Once implemented in a company, an Environmental Management Systems (EMS) determines the environmental impact of the company activities, defining strategies, targets, and measures to prevent or minimize negative impacts. It also preserves the previously acquired knowledge and experience. The EMS achieves a continuous improvement in environmental protection, complying with existing environmental regulations. The Environmental Management cycle can be seen in Fig. 9.13.

Among the EMS advantages, the following can be mentioned:

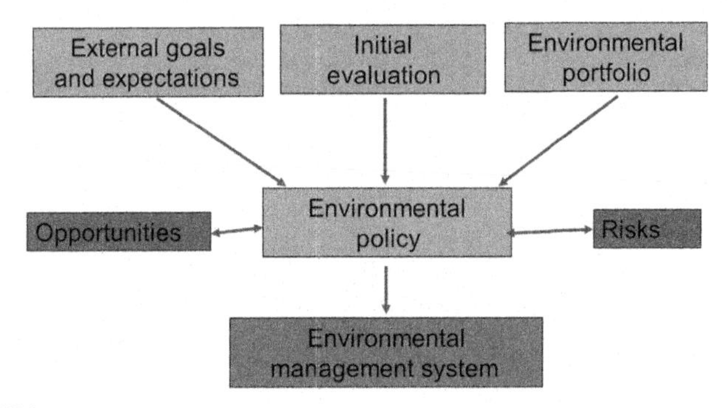

Fig. 9.12 EMS inputs.

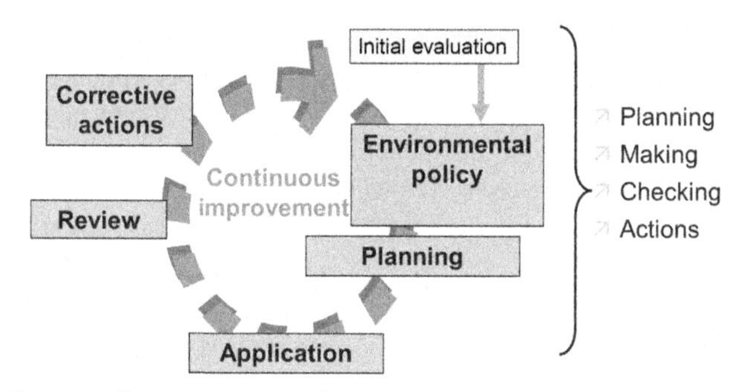

Fig. 9.13 Environmental management cycle.

- Systematize the environmental activities
- Minimize risks thanks to legal normative compliance
- Cost savings
- Improve the Corporate Image in front of the regulators, the customers, and the general public
- Attract and retain customers and increase stock market value
- Maintain long-term competitiveness

The most widely recognized EMS regulation is the ISO *(International Organization for Standardization*), which is an independent nongovernmental organization based in Geneva, with 164 Member States. ISO has prepared almost 20,000 standards of voluntary application, among them:

ISO 9000—quality
ISO 14000—environment
ISO 20121—event sustainable management
ISO 26000—social responsibility
ISO 50000—energetic management

An EMS, following ISO 14001 rules, requires (see Figs. 9.14 and 9.15):

- Compromise of legal compliance
- Compromise of continuous improvement

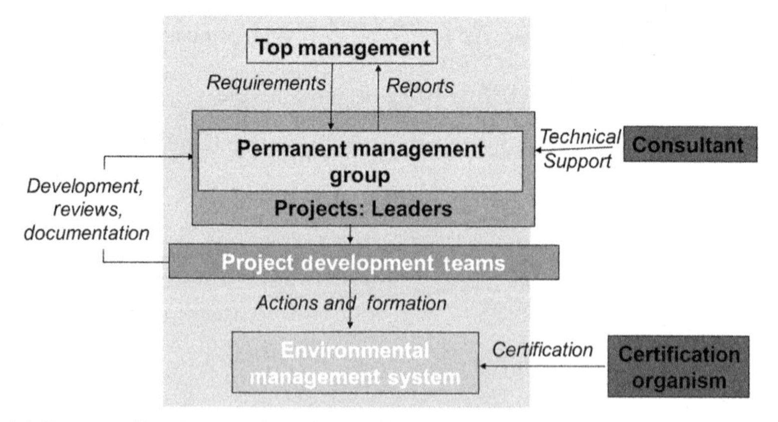

Fig. 9.14 Process of implementation of an EMS.

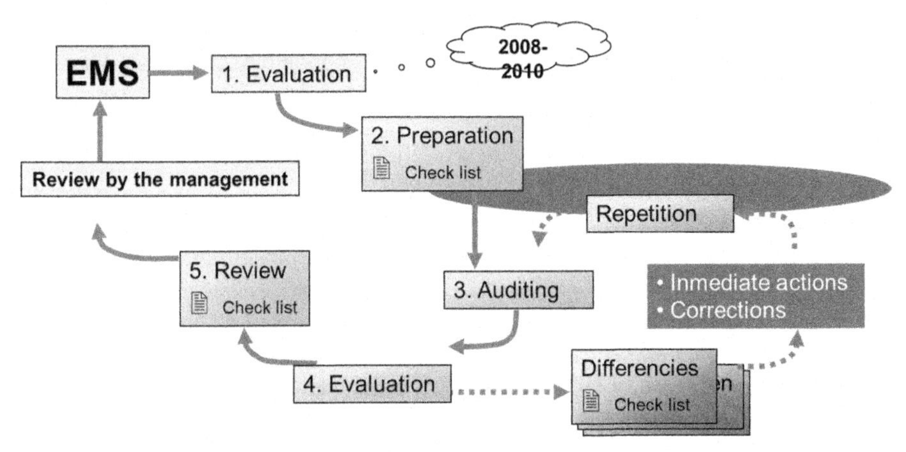

Fig. 9.15 ISO 14001 EMS implementation steps.

- Declaration of environmental relevancy
- Public announcement of the main strategies
- Communication and Cooperation Plan

ISO 14001 certification is an internationally recognized EMS that provides the basis for internal management. It is a guarantee of an external and independent system review and it is also an important image and Public Relations element.

Some system basic elements to be considered when implementing an EMS are:

- Knowledge of the applicable regulation
- Selection of the most relevant environmental impacts
- Definition of environmental impact indexes
- Establish the reduction goals
- Reduction plans and their responsible ones
- Formation and Communication Systems, external and internal

Summarizing, an EMS is a tool for the environmental impact control of the company (airport, airline, etc.) or of one activity in the company (maintenance, handling, etc.). The application structure is specific of each company or included activity. To be successful, an EMS needs to be fully understood, applied, and checked. It is a long-term philosophy and not a command. A good environmental management is not a guarantee for the good financial health of the company, the employees' satisfaction, the shareholders' approval, or any growth opportunities... but it certainly may help to achieve all of them.

Nowadays, all industries around air transport pay special attention to environmental management. It is the case of the aircraft manufacturers. Boeing for instance publishes that "certification to the internationally recognized ISO 14001 environmental management standard has strengthened our ability to meet our aggressive environmental targets. Sites in Australia, Canada, the United Kingdom, Malaysia and the United States are certified to the global ISO 14001 standard. This includes all Boeing's major manufacturing sites as well as sites devoted to engineering, business development, customer support and administrative functions. Currently we have more than 50 Boeing locations certified, with more locations pursuing certification." Equally Airbus announces that "Airbus has set up and maintains an EMS that aims at reducing the environmental impacts of both our sites and products operations and is compliant with the ISO 14001 certification."

9.8 Sustainability

A widely accepted definition of Sustainability is *"development that meets the needs of the present without compromising the ability of future generations to meet their own needs" (Our common future,* a 1987 report to the United Nations by the *World Commission on Environment and Development,* led by *Gro H. Brundtland).* This definition was confirmed in 1992, during the United Nations Conference on environment and development, held in Rio de Janeiro, and in the following *Earth Summits* in Johannesburg (2002) and Rio de Janeiro (2012).

United Nations develops its sustainability general policies through a permanent organization, the *Commission on Sustainable Development.* The most practical transportation issues are dealt with by the *International Transport Forum (ITF).*

From a different perspective, transport policy socioeconomic matters are discussed in the *Economic and Social Council* (ECOSOC). UNECE is integrated by 56 States in Europe, Central Asia, and North America. UNECE works for the application of sustainable development policies, including common regulations adapted to the international transport needs.

Another approach to the sustainability problem is proposed by the Club of Rome, an organization created as an NGO in 1968. Its first report, *The limits of Growth,* published in 1972, discussed about the relationship between demographic growth and consumption (prepared by a MIT team, led by *Donella Meadows).* In 2012 The Club of Rome published *"2052; A Global Forecast for the next Forty years,"* centered on Climatic Change.

The Club of Rome and other ecologist organizations try to compare the resource consumption levels with the Earth capacity to generate those resources. The calculation unit is the amount of resources generated during 1 year versus the consumption of resources by the world population. In 1993 Level 1 (consumption equal to the annual production) was reached in October, in 2003 in September, and in 2013 in August, according to the NGO *Footprint Network.*

It is different talking about sustainable transport development and transport for a sustainable development. Transport, being a derivative activity and not an end in itself, serves to a large number of primary activities (trade, business, culture, leisure, etc.), which must be sustainable. Some activities are improved and/or grow, thanks to the value added by transport, but others are directly created due to the existence of convenient transport services.

On the other hand, the definition of Sustainable Transport would be the capacity to satisfy social needs (mobility and accessibility for people and goods; trade and communication; social and cultural relationship) without sacrificing other essential values (human/social; ecologic) in the present time without compromising the ability of future generations to satisfy their own needs.

The European Union in their second Treaty of Maastrich established the instruments for a sustainable transport:

- General economic incentives (ecologic taxes)
- Specific incentives per transport mode or effect
- Specific regulations per transport mode or effect
- Complementary transport coordination policies

Transport policies define the mode to be used through four elements:

- Technical regulation (i.e., different width in railway tracks)
- Commercial regulation (i.e., traffic rights concessions)
- Infrastructure development (i.e., roads versus highways)
- Cost externalization (i.e., payment for infrastructure use)

Sustainable development has to rightly consider three elements: social, economic, and environmental effects, in five action areas:

- Physical accessibility (physical barriers to the transport mode use, as distance or physical capability)
- Economic accessibility (price barriers)
- Safety (statistical possibility of suffering an accident)
- Security (statistical possibility of suffering a menace, attack, or robbery)
- Environment (local and global environmental impacts)

In March 2012 the air transport sector (airlines, airports, air navigation service providers, and aircraft and engine manufacturers) unanimously signed a document: *Toward sustainable aviation*, explaining their targets in this field and the areas in which governments' participation is needed to reach the expected results. The goals were:

- Recognize that Air Transport as an economic growth catalyst, increasing trade, personal mobility, and tourism through greater connectivity.
- Creation of high value-added tasks, keeping a high research & development investment level.
- Do not increase CO_2 net emissions from the year 2020 level (CNG Carbon Neutral Growth) and reduce them 50% in 2050, compared to the 2005 year.

Government participation:

- Internationally coordinated research investments
- Urgently improve the world ATC system efficiency
- Incentives to the use of renewable alternative energies
- Airport infrastructure development
- Environmental regulations for encouraging the environmental impact reduction
- Reach a worldwide agreement in ICAO to reduce international civil aviation emissions

In 2015, during the COP/21 in Paris 195 States agreed on reducing GHG emissions as soon as possible and keeping Global Warming well below 2°C. In order to be mandatory, the agreement should be ratified by at least 55 States, representing not <55% of the GHG emissions. A ceremony to open the ratification period was held in April 2016 in New York. Interestingly, aviation is not specifically mentioned in the agreement.

Efficiency regulation and certification

10

Commercial aircraft energetic efficiency has always been a paramount target for the manufacturers, but the results have been evaluated by the individual operators, without any CAA intervention. Comparisons between different products and historical trends showed clear progress but quantitative assessment was dependent on the type of operator that was making the numbers. This situation has changed in the last decade, when ICAO has made a big effort in achieving a more neutral and verifiable procedure.

This last chapter describes the development of two new ICAO regulations: the ICAO CO_2 certification standard, applicable to new commercial aircraft types starting in 2020 and in-production aircraft some years after, and the emissions offsetting system, known as CORSIA, which will be applied to keep net international aviation emissions at 2020 level, following the CNG target explained in the previous chapter.

As already explained, there are many possible ways to measure air transport energetic efficiency. ICAO uses the standard RTK/fuel (kg) for the establishment of its policy targets and the goals of individual States Action Plans. In this way, the metric covers the whole system, including aircraft, operations, and infrastructure. The isolation of those factors in order to know which part of the progress can be attributed to each one is not an easy task.

At the same time, ICAO has a long historic trajectory of certifying different environmental aircraft features, like noise or LAQ engine emissions, settling limits for new aircraft/engine types. The limits are becoming more stringent as technology moves forward. The philosophy behind this policy is not to force manufacturers to perform extraordinary efforts into exploring very much advanced technologies, but to confirm that everybody is up to date in environmental design features and there is no step backwards. Improvements in one of the environmental factors must not be made at the expense of deteriorating other different areas. When ICAO approves a new standard, a sound analysis has to demonstrate that the adopted measure is technically feasible, economically reasonable, and environmentally effective.

Any new commercial aircraft type or commercial jet engine has to demonstrate compliance with the ICAO Annex 16 standards, using detailed certification procedures, before a Type Certificate is granted. Part 1 of the Annex contains noise aspects while engine LAQ emissions are dealt with in Part 2. The certification procedures determine the tests to be performed in order to demonstrate standard compliance. For the noise, there are three measurement points recording approach, takeoff, and lateral noise. LAQ emissions are measured in engine test cells, simulating all the approach, landing, taxi-in, taxi-out, takeoff, and initial climb made by the aircraft in a normal operation, below 3000 ft. over the airport. There may be discussions with respect to how representative are the measurement conditions compared with the actual operation, but the technical details are unambiguous and it is generally accepted that the obtained values are representative of the environmental technology applied in the aircraft/engine design and manufacturing.

Energy Efficiency in Air Transportation. https://doi.org/10.1016/B978-0-12-812581-6.00010-7

The idea of adding a fuel efficiency certification requirement to foster advanced technology application has been around the industry during the last 30 years. A number of arguments support that action: the importance of fuel consumption efficiency for economic and environmental reasons, the need of making clearer the progress made in the last decades, the possibility of differentiating aircraft models by a transparent efficiency metric, and the advantages of having a smart tool for future regulation, in terms both technical and economic.

Looking around into other sectors, ground transportation has offered officially tested fuel efficiency and CO_2 values, included in the documentation supplied to any prospective car buyer. The values are determine by test performed in circuits reproducing urban or highway traffic conditions, providing a range of values that may give an idea of the expected consumption of the vehicle. It was surprising, at least, that a high-technology sector like commercial aviation was not able to offer a similar level of information to the public.

However, this type of certification for commercial aircraft presents additional and more difficult challenges than noise and LAQ emissions. First, the definition of the subject to be measured is more complicated, as the efficiency, in terms of RTK/fuel kg changes, depends on the route features where the aircraft is flying, the type of aircraft architecture for that operation, and the standard procedures of the airline operating that route. Long versus short range, turbofan and turboprop, single aisle or wide body, passenger aircraft or freighter, and the combination of all of them make difficult to create a common set of conditions where to calculate energy efficiency in an unbiased way.

Commercial aircraft performance has a number of certificated values included in official Performance and Operations Manuals. However, those data reflect the expected performance of the aircraft in a certain determined condition, with a substantial safety margin included, and cannot be taken as a comparative standard. At the same time, fuel efficiency is a magnitude extremely important at the moment of selecting the composition of a new fleet. Airlines devote a huge amount of efforts and resources in the comparison of different aircraft models, all of them being promoted as the most efficient ones by their manufacturers. The availability of an official, certificated figure indicating that aircraft A is X% more efficient than its competitor aircraft B could made a big difference in the marketing chances of those two models. In consequence, manufacturers were not enthusiastic with the idea.

10.1 The CO_2 certification requirement

As indicated in the previous Chapter 9, ICAO is leading the actions for mitigating the effects of international aviation activities in climate change, following the Kyoto Protocol Agreement (1997) that entered into force in February 2005. Domestic flight emissions were included in national inventories but, from a technological point of view, there is no difference between the aircraft fleet of international and domestic markets and adopted actions will have repercussions in both areas.

ICAO actions started with the petition of a wide range study to the Intergovernmental Panel on Climate Change (IPCC), covering the effects of the aviation activities on the atmosphere, published in 1999. This document (described in Chapter 9) contains a rigorous analysis of the civil aviation impact in the atmospheric conditions, based on the 1992 traffic situation, and presented different scenarios going up to 2050.

After receiving that study, during the 2001–04 period, the ICAO Committee on Aviation Environmental Protection (CAEP) performed a detailed study on the possible MBMs to be adopted, divided into four groups: voluntary commitments, charges, taxes, and Emissions Trading System (ETS). The results showed a clear advantage for the latest and, in the next ICAO cycle 2004–07, CAEP-approved guidelines to facilitate the implantation of a worldwide ETS. However, the 2007 ICAO Assembly failed to get an agreement on this point. Subsequently, the European Union put into force an aviation ETS for the flights between two EU airports, starting in January 2012, and indicating its intention to expand it to cover all the flights in its airspace unless a similar or equivalent worldwide mechanism was approved by ICAO.

In the 2010 37th Assembly, ICAO approved a number of measures to limit CO_2 emissions, including National Action Plans for the Member States with >1% of the total international RTKs, operational improvements, and the goal of developing procedures to certify the fuel efficiency of new aircraft types, while an alternative MBM, acceptable for the great majority of Member States, is discussed. Following the work division within ICAO, these tasks went into CAEP domain and its Working Group 3 (Emissions) job. In this way, the CO_2 certification was officially included in the colloquially called "basket of measures" addressed to control the air transport contribution to climate change.

A specific Task Group was created to elaborate the CO_2 certification procedure first and the stringency levels to be applied later. The first meeting of the CO_2 TG was held in May 2010 and the final proposal was agreed in October 2012, to be finally endorsed by the 38th Assembly in autumn 2013, being published as ICAO Circular 337 *Agreed Certification Requirements for Aeroplane CO_2 Emissions Standards* in the following year.

10.2 Developing of the metric

The first part of the work was addressed to select a metric representing the fuel efficiency of the technology applied to each aircraft type. This was a difficult task because, in addition of the high representability level, the selected magnitudes should comply with the requisites of being certifiable without imposing excessive economic or time penalties to the manufacturers and the CAAs in charge of the certification. After some lengthy discussions, six key criteria (KC) were defined:

KC1. *General.* The certification standard must not compromise safety levels. The CO_2 certification requirement should be airplane performance-based, should reflect CO_2 emissions at the airplane level. It should also allow for the differentiation of products with different generations of CO_2 reduction technologies and should aim to be independent of airplane purpose or utilization. The certification requirement should decouple effects of fuel performance from airplane performance.

KC2. *Effective.* Improvements observed via the CO_2 certification requirement should correlate with reduction of CO_2 emissions at the airplane level as demonstrated by procedures,

which are relevant to day-to-day operations. It should also take into account fundamental airplane design elements and capabilities (such as distance traveled and what is transported).
KC3. *Objective*. The certification requirement should be objective and therefore needs to be based on certified airplane parameters and/or currently noncertified parameters. The parameters that compose the metric should be easily measurable at the certification stage, or derived from engineering data. The certification requirement should consider the industry standard practices of measurement and adjustment.
KC4. *Robust*. The metric should be robust in order to minimize the potential for unintended system and airplane design consequences, to limit interdependencies, and to limit any influence on other standards. To the extent practicable, the certification requirement should be fair across the set of stakeholders, such as manufacturers and operators.
KC5. *Reasonable*. The certification requirement should not require an inappropriate level of resources on the part of National Airworthiness Authorities and manufacturers to implement. If the certification requirement requires the certification of additional parameter(s) compared to existing practices, the implications (such as technical feasibility and economic reasonableness) should be evaluated.
KC6. *Open*: The output should be explainable to the general public.

Those key criteria were complemented by a set of three high-level principles:

- Within the "basket of measures," an aircraft CO_2 standard should focus on reducing CO_2 emissions through integration of fuel efficiency technologies into airplane-type designs.
- Aim to design a metric system (metric/correlating parameter/test points), which could permit transport capability neutrality at a system level when stringency is applied based on this metric system.
- Aim for equitable recognition of fuel efficiency improvement technologies in an aircraft-type design.

Since the beginning of discussions, there were two basic options considered: a metric focused on fuel burned per RTK or other looking at specific aircraft range (SAR) values. ICAO standards, including all Annex 16 noise and emissions, are focused on making sure the incorporation of the best available technologies developed at the time of the type certificate application. In any case, the new standards have to be technically feasible, economically reasonable, and environmentally beneficial. Then, all these three features need to be present in the selected metric.

The first option was in line with the ICAO way of measuring fuel efficiency goals and was more intuitive as a metric of transport (represented by tonne-kilometer) related to the amount of fuel needed to produce it. However, it was more difficult to certify, because the definition of a full-flight condition in order to measure the total fuel consumption presents great difficulties due to the high number of involved flight parameters and their extreme variability, a factor very determinant on certification procedures that need good levels of standardization.

The use of specific range (kilometers per amount of fuel) does not allow a direct perception of the efficiency, because it misses a magnitude of what is transported. An additional element to include this last factor should to be added. The measure of specific range in an instrumented test aircraft is easier to measure and standardize and provides a good proxy of the propulsion efficiency and the aerodynamic refinements of the aircraft design. The structural efficiency is not taken into account and requires an additional factor with a relationship between the aircraft design and the payload capacity.

The comparison of the payload efficiency among different generic configurations of aircraft so different as turboprops, regional jets, short-medium, and long-range commercial airliners is a formidable challenge. An additional difficulty comes from the fact of including freighters in the group, the payload of which used to be greater than the one corresponding to the same type of aircraft in passenger version.

The adopted solution was to use a geometrical parameter instead of a weight one. This was named reference geometric factor (RGF) and represented the floor surface available to receive payload in the cabin(s) of the aircraft. The geometric definition is complex because it has to cover single- and double-deck aircraft (e.g., A330/B777 vs. A380/B747), narrow and wide bodies (MD80/B757 vs. B787/A350), jets and turboprops (CRJ700/EJ170 vs. Dash-8/ATR72), and fuselages with circular (like A320 family) and double oval (B737 family) sections. At the end, the adopted definition was:

(a) For airplanes with a single deck, the area of a surface bounded by the maximum width of the fuselage outer mold line (OML) projected to a flat plane parallel with the main deck floor.
(b) For airplanes with an upper deck, it is the sum of the area of a surface bounded by the maximum width of the fuselage outer mold line projected to a flat plane parallel with the main deck floor, and the area of a surface bounded by the maximum width of the fuselage at or above the upper deck floor projected to a flat plane parallel with the upper deck floor.

The limits of the surface to be computed in the RGF include all pressurized space on the main upper deck including aisles, assist spaces, passage ways, stairwells, and areas that can accept cargo and auxiliary fuel containers. It does not include permanent integrated fuel tanks within the cabin or any unpressurised fairings, nor crew rest/work areas or cargo areas, which are not on the main upper deck (e.g., "loft" or under floor areas).

The rear boundary is always the rear pressure bulkhead. In a similar way, the forward boundary is the forward pressure bulkhead except for the cockpit crew zone (the area of the airplane designated solely for crew use). If the aircraft has a cockpit door, this will be the forward limit, always considering the cockpit door option giving the largest RGF. If there is no such cockpit door, such as airplanes capable of single-pilot operation, the forward limit is given by passenger access. Areas not accessible to passengers in all interior arrangements shall be excluded from RGF calculation.

Figs. 10.1 and 10.2 provide a notional graphic representation of those concepts.

For the SAR calculations, the idea was to take the average of three points, placed at the beginning, in the middle, and at the end of the cruise phase, as seen in Fig. 10.3. The weight corresponding to each one of the three points is as follows:

- High gross weight: 92% MTOM
- Mid gross mass: Average of high and low gross mass
- Low gross mass: $(0.45 \times \text{MTOM}) + (0.63 \times \text{MTOM}^{0.924})$

MTOM being the maximum certified take off mass in kilograms.

It is necessary to recall than, for this calculation and the subsequent flight test, the cruise is always flown in the optimum conditions and, consequently, the aircraft is not

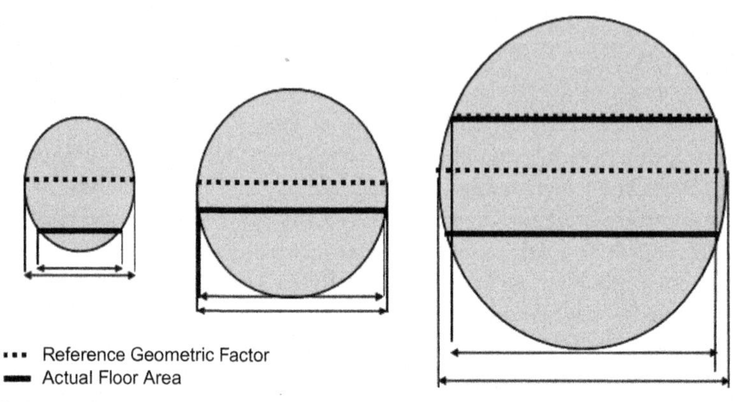

Fig. 10.1 Determination of the level to calculate RGF, depending on the fuselage section (*ICAO: Circular 337*).

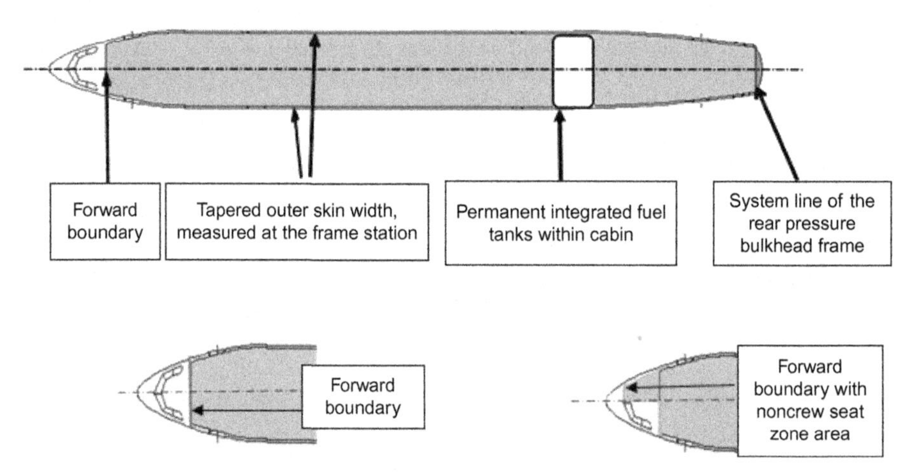

Fig. 10.2 Limits of the horizontal surface for RGF calculation (*ICAO: Circular 337*).

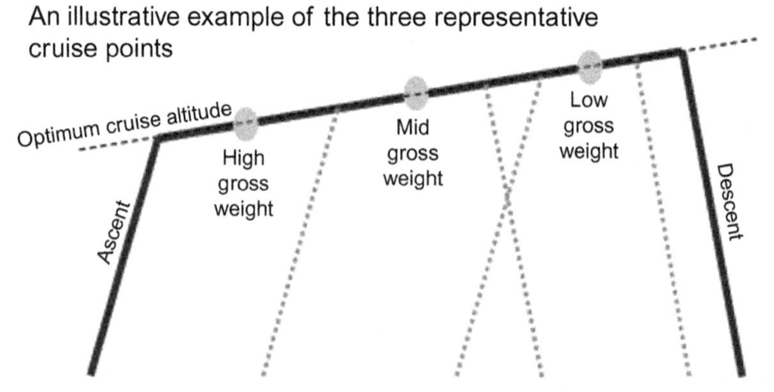

Fig. 10.3 Maximum range profile for efficiency certification (*Dickson, Neil: Global emissions technology*).

operating at constant altitude, but increases flight level in a continuous way, as it loses the weight of the consumed fuel. A rough description of the theoretical procedure is given in Fig. 10.3.

The final formulation of the metric value (MV) is:

$$MV = (1/SAR)_{AVG} / RGF^{0.24} \tag{10.1}$$

with basic dimension of kilogram of fuel per kilometer. The initial coverture of the formula was thought for subsonic jet airplanes with MTOM over 5700kg and propeller-driven airplanes with MTOM over 8618kg, with application for Type Certificate submitted on or after 1st of January 2020.

An analysis of the present aircraft types MV showed a reasonable relationship with MTOM and this magnitude was adopted as the right correlating parameter for the MV. Each aircraft type was now represented by a point with MV and MTOM coordinates.

During the discussions to approve the standard limits to be complied with by those MVs, the application of the same limits to models in production before the deadline of the regulation and to modifications of those in production (In *P*) model that was the subject of modifications not requiring a new Type Certificate but producing a not-negligible change in their energetic efficiency was debated. These two issues were widely debated during the standard definition discussions held in the 2013–16 period.

10.3 The standard approval

After 2013 Assembly approval of the metric, the next step was to decide the limits to be accomplished by future airplane types. Within CAEP, regulatory standards have a well-defined working path. First were the corresponding technical Working Group, in this case Working Group 3 (emissions), analyses of the situation of airplane types in production or in development, calculating the corresponding Metric value for each one of them. Then, the shape of the limit line is decided and a number of possible stringency options is agreed. Those options are evaluated by the Forecasting and Economic Support Group (FESG) in terms of cost and environmental benefits for the aviation industry and the results are submitted to the CAEP. The group prepares a proposal for the Assembly, either selecting the best candidate option or two or three possible choices.

The CO_2 metric value limit shape was highly complicated by the fact that there were no previous references for this analysis. To solve this point, a reference line in the graphic representation MV/MTOM was created in such a way that all the In *P* aircraft will be in compliance with that limit. The line was impossible to fit straightforward and had to introduce a kink at MTOM = 60 ton to reflect the somewhat different features of small aircraft.

With respect to the Adapted Reference Line (ARL), up to 10 different stringency grades were created, applying fixed percentages of MV. Table 10.1 shows the features of the 10 different Stringency Options (SO) to be evaluated by FESG, with the MV at the 60-ton kink point and the percentage of aircraft models, presently in production, that failed to comply with them.

Table 10.1 Stringency options and their impact in the existing aircraft types (*ICAO: CAEP10 report*)

CO_2 stringency option	% to ARL at 60t MTOM	MV at 60t MTOM	Impact rate
1	−20.0	0.8734	7%
2	24.0	0.8297	13%
3	27.0	0.7970	24%
4	28.5	0.7806	32%
5	30.0	0.7642	41%
6	31.5	0.7479	50%
7	33.0	0.7315	57%
8	34.3	0.7173	60%
9	37.5	0.6823	79%
10	40.4	0.6507	99%

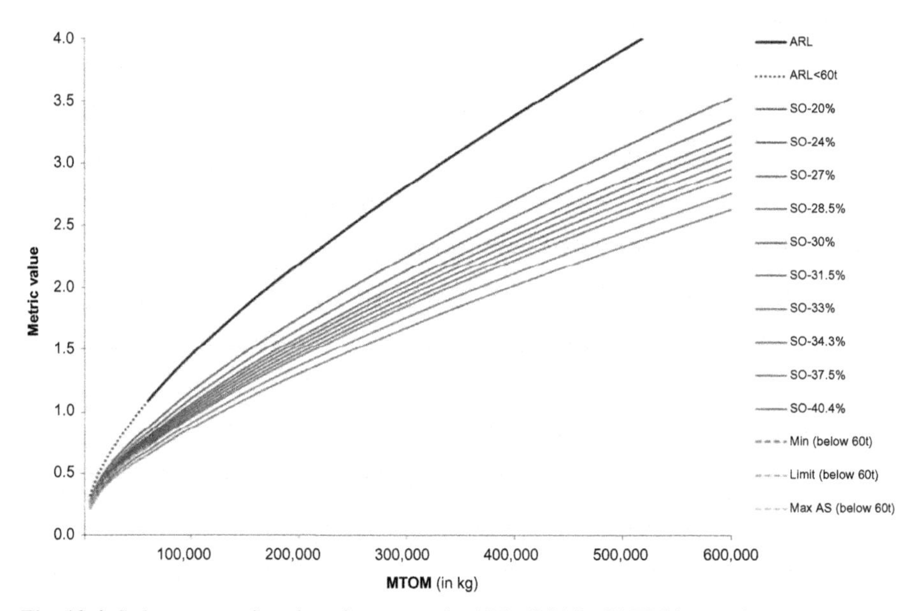

Fig. 10.4 Stringency options in reference to the ARL (*ICAO: CAEP10 report*).

Their representation in the standard Metric Value—MTOM can be seen in Fig. 10.4. Fig. 10.5 changes the scale and includes only aircraft with MTOM up to 100 ton in order to underline the effect of the kink point at 60 ton.

The evaluation scheme can be seen in Fig. 10.6. It is a complex system, which has a fleet growth and replacement model as a base. Under a traffic growth assumption, World fleet models are classified by size and range and distributed in pairs. Each stringency option asks for replacement of noncomplying models, replaced by aircraft types in the same category. At the same time, some market responses are assumed because some noncomplying existing models might be modified, at a certain investment amount, to improve their performance and become within the stringency.

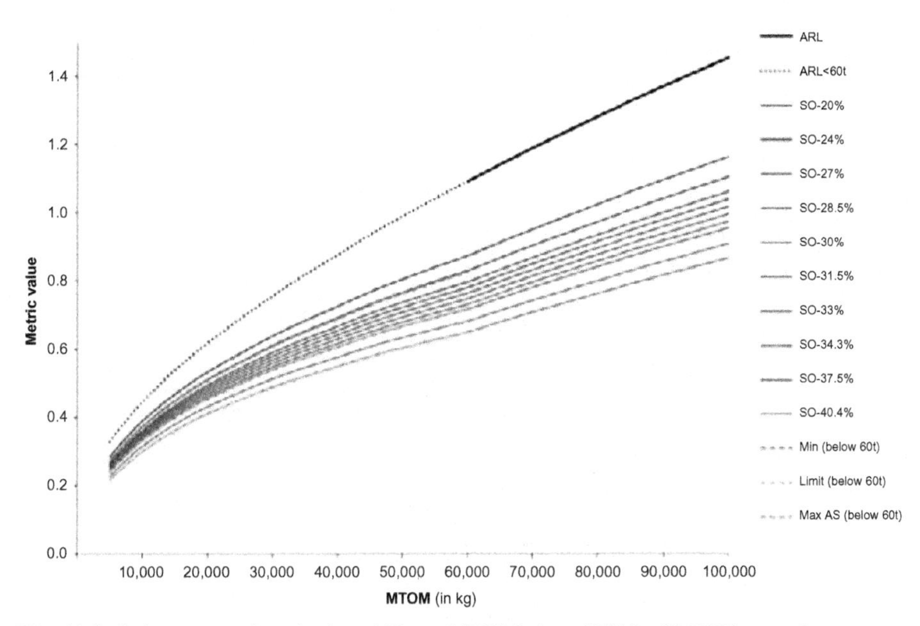

Fig. 10.5 Stringency options in the <100-ton MTOM class (*ICAO: CAEP10 report*).

Another important element in the analysis is the consideration of interdependencies, because CO_2 standard compliance must not compromise to meet other environmental regulations, the most relevant for this feature the new Annex 16 Part 1 Chapter 14 maximum noise levels, entering into force by 2018–20 (depending on the MTOM of the affected aircraft), almost at the same time than the CO_2 standard. In the case of NO_x the basic assumption was a relationship 1:1 between fuel consumption and LTO and NO_x in cruise. Then, fuel consumption reduction was parallel to NO_x reduction.

The results of comparing the cost increase and the emissions savings of each stringency option with a reference level, corresponding to the hypothesis of no ICAO action, were classified in terms of cost per CO_2 ton reduction. However, the number of possible alternatives was very high, because each stringency level might be applied only to new types, to new and In *P* types at the same time or in different dates, or different SO might be used for new types and In *P* types in the same or different dates.

With respect to the In *P*, three options were contemplated:

- Option 1 represents a production cut-off for all In *P* airplane types if they have not been certified to the CO_2 standard by this date.
- Option 2 represents applicability being triggered only if an application for a design change for new in-production airplanes exceeds a specific CO_2 change criteria.
- Option 3 represents a hybrid approach, where Option 2 is active for a period of time and is subsequently followed by Option 1.

The political factor appeared when SO application results were computed by different manufacturers. It was evident that a number of models, originating in the former USSR and presently being manufactured in Russia and Ukraine, could not meet most of the SOs (there were some failing even SO 1) and were unsuitable to

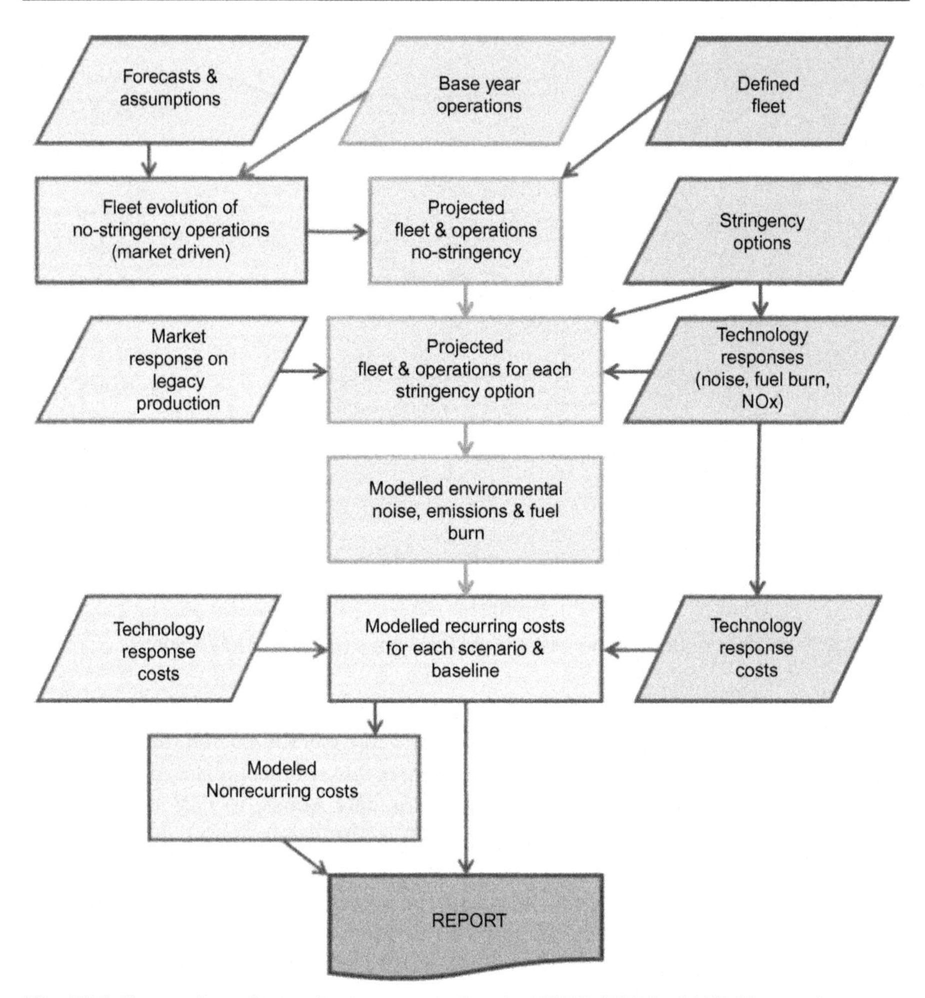

Fig. 10.6 Economic analysis of stringency options by FESG (*ICAO: CAEP10 report*).

be modified. At the end, an agreement was reached allowing an exemption for a limited number of those Russian and Ukrainian models, representing a very small amount of emissions. The present political incertitude made very difficult any accurate assessment about the future production of those types, but it is believed to be very low.

The final agreement, proposed by CAEP and endorsed by ICAO 2016 General Assembly, was based on four points, covering both New Types and In *P* aircraft:

- for New-Type airplanes >60-ton MTOM, a stringency level of SO 8.5 with an applicability date of 2020;
- for New-Type airplanes less than or equal to 60-ton MTOM, a stringency level of SO 5 with an applicability date of 2020, and a later applicability date of 2023 for airplane-type designs with a passenger-seating capacity of equal to or <19 seats;

- for in-production airplanes >60-ton MTOM, a stringency level of SO 7 with applicability trigger Option 3, an applicability date of 2023, and production cut-off 2028;
- for in-production airplanes less than or equal to 60-ton MTOM, a stringency level of SO 3 with applicability trigger Option 3, an applicability date of 2023, and production cut-off of 2028.

In its March 2017 meeting, the ICAO Council issued the definitive text for the CO_2 certification process and the standard to be included as Part 3 of the next Annex 16 edition. As other standards of this Annex, there will be an open database with the aircraft types MV as soon as they are being certified.

The effects of the new standard on global aviation CO_2 emissions are certainly subject of debate. Fuel consumption being a very important factor in airline fleet decisions, the existence of a certified figure, the MV value, provides a valuable orientation to aircraft evaluators. Even accepting that MV is not directly representative of actual consumption, it seems likely that manufacturers will work hard to classify their products in good positions within commercial aircraft MV ranking. A recent analysis by the International Council on Clean Transportation (ICCT) says that the standard will on average require a 4% reduction in the cruise fuel consumption of new aircraft starting in 2028 compared with 2015 deliveries, with the actual reductions ranging from 0 to 11%, depending on the MTOM of the aircraft.

The results of the study are very different for new types and In *P* models. Table 10.2 shows that, according to the available information, modern types (starting with types certified in 2011, as Boeing B747-8 and going up to Boeing B777-9 that will enter into service in 2020) will be little directly affected by the new rules, although there is always the competitive effect among the manufacturers.

The situation is very different when In *P* aircraft types suffer the new standard. Emissions reductions are required in all categories, as seen in Table 10.3, even disregarding the low-production models to be exempted. Large aircraft and freighters will be those most affected if they will continue in production after 2028.

Following the trend adopted with other standards, the technological base of SO levels will be revisited periodically. It is worthwhile to notice that the standard aircraft database, including all aircraft in production and some of the models in development, the data of which were delivered by their manufacturers and verified by their respective

Table 10.2 Estimated standard effect on modern aircraft types (*ICCT: International Civil Aviation Organization's CO_2 standard for new aircraft*)

Aircraft type	Standard application	MTOM (tonnes)	MV (kg/km)	Required MV	% of reduction
Very large aircraft	2020	>350	2.75	2.51	10
Twin aisle	2020	120–350	1.51	1.68	0
Single aisle	2020	60–120	0.76	0.82	0
Regional jets	2020	13.5–60	0.55	0.66	0
Business jets	2023	<60, <19 seats>	0.46	0.58	0

Table 10.3 Estimated MV reduction required by the standard application for In *P* aircraft (*ICCT: International Civil Aviation Organization's CO_2 standard for new aircraft*)

Aircraft category	MTOM (tonnes)	MV2015 worst	MV 2015 average	MV2028 target	% of reduction worst a/c	% of reduction average a/c
Very large a/c	>350	2.95	2.93	2.62	11	10
Twin aisle	120–350	1.88	1.70	1.75	7	0
Single aisle	60–120	0.94	0.91	0.86	9	6
Regional jets	13.5–60	0.71	0.69	0.68	3	0
Business jets	<60, <19 seats	0.64	0.56	0.61	6	0
Freighters	n/a	2.13	2.06	1.92	10	7
Average		1.59	1.49	1.46	8	4

CAA, will be public and open to all the interested parties. This covers more or less aircraft types with entry into service dates up to 2020. The advances in technology and its application to new models may give a better idea of the consequences of the approved resolution. The first technology review has been initiated in 2016 and will be concluded in 2019.

10.4 Offsetting systems

As the effect of CO_2 emissions on the climate change depends only on the substance weight emitted and is independent of the place of emission and the emissions source, the United Nations Framework Convention on Climate Change (UNFCCC) allows compensation of emissions from activity sector to others under a certain number of conditions.

In the third meeting of the Conference of the Parties (COP), held at Kyoto in December 1997, developed States received targets to reduce their Green House Gas (GHG) emissions. In addition to the direct reduction systems applying technology or modulating the amount of activity in some areas, States may "buy" emissions credits using three market mechanisms:

- International Trading Systems, in which a emissions allowances market is created and participants may buy or sell them, according to their needs, depending on whether they are over or below a pre-established emissions target. This is the system used in the European Union since 2005 for a number of industrial sectors and, since 2012, for the civil flights between two EU airports.
- Clean Development Mechanism (CDM), if a State with a GHG target finance program to reduce GHG emissions in other countries that were not affected by the targets.
- Joint Implementation (JI), through joint international initiatives with other States included in the GHG targets.

Domestic aviation was included in GHG inventories of each State, but international aviation, as international maritime traffic, was not allocated to the national inventories, but left it to the authority of the respective international organizations: ICAO and its counterpart the International Maritime Organization (IMO). In both cases, the general rules and mechanisms were valid and, for example, ICAO approved in 2007 a very comprehensive guideline document on the application conditions for those States wishing to implement an Emissions Trading System.

The interest of compensation schemes for aviation, either CDM or JI, is based in the idea that aeronautics is a very technologically advanced sector, where fuel efficiency is a key element from both environmental and economical points of view. Therefore, the "easy" wins have already been achieved and going further would require high investments per unit of saved emissions, while there are other areas in which the same CO_2 reduction would come at a much lower price. As a consequence, it might be more efficient to invest in these other sectors if the emissions savings could be accounted for the aviation sectors.

This approach gives the aviation decision makers three options to comply with any emissions target: invest into improving the efficiency of its operation (operations, fleet replacement, load factors, maintenance), reduce growth to the level required by its emissions target, or invest in compensatory or offsetting methods through other activities.

The best known system, allowing passengers to offset the CO_2 emitted by their flights, was introduced by many airlines in their booking systems. At the time of purchasing a ticket, the application shows the amount of CO_2 corresponding to a single passenger in that trip and asks the customer whether he/she wishes to offset it, paying an additional amount. The weight of CO_2 is computed through an emission calculator model that has into account the type of aircraft and the distance of the flight. The offset price is the cost of reducing that amount of emissions by spending the money in a specific environmental program supported by the airline.

Offsetting systems have a number of advantages: they are voluntary and relatively simple to be understood, give relevancy to the individual participation of the travelers, and allow the incorporation of other partners, like NGOs. However, they have several features that reduce their credibility and have diminished their propagation since their inception.

From the administrative point of view, the activities focused on emissions reduction need to be adequately verified and monitored. That requires an official certification with a cost level that makes it nonviable for small programs. IATA is trying to overcome this problem by grouping small- and medium-sized airline programs in big packages capable of financing big compensatory programs.

The credibility of the compensatory programs is always important as the customer needs to be sure of where his money goes and being confident on the right purpose of those activities. Reforestation and natural landscape protection used to be the most attractive programs. A second and non-negligible element of mistrust is the different values shown by different calculators. ICAO has tried to unify them in an official system that may be addressed directly in the ICAO web page and it is now used by the majority of the airlines.

A third factor is the position of the offset tool within the payment chain. In present times, most airline tickets offers the possibility of different additional payments on the basic fare, like seat selection, additional baggage, preferential boarding, additional insurance, and other ancillaries. When the customer has covered some of these possibilities, carbon offset becomes an extra cost, not very appealing. Some NGOs are asking the airlines for an opt-out system, where the customer has to explicitly say that CO_2 offset is rejected, instead of the present opt-in, in which offsetting will needs to be indicated.

The accounting of all those compensations in a global worldwide scenarios is not easy if we try to avoid duplications. For example, a tree plantation paid by this type of voluntary airline programs cannot be computed as a forestry operation, as the CO_2 absorption is already computed in the airline net carbon footprint. For being even more accurate, the new trees absorption of CO_2 should be reduced by the existing absorption of the field where the trees are now planted.

Some entities or environmentally minded companies automatically offset the carbon emissions of their employee trips, but the total volume of those programs has been very modest until the recent approval of the CORSIA by ICAO.

10.5 The CORSIA system

As it was recognized that air transport growth has historically surpassed efficiency improvements, leading to a fuel consumption increase situation, economical measures are considered essential to achieve ICAO target of carbon neutral growth after 2020. The underlying idea is that worldwide international aviation should become active in the carbon allowance market and compensate by buying those allowances the amount of emissions exceeding the 2020 levels.

The most important obstacle for a worldwide adoption of that goal was to respect two basic elements inserted in the general ICAO philosophy:

- Common But Differentiated Responsibilities (CBDR)
- Special circumstances and Respective Capabilities (SCSR)

The CBDR idea takes on board all the ICAO member States, stating that CO_2 reduction is a common task but, at the same time, recognizes that responsibility levels are different, because the amount of emissions by their aviation is widely different and, in addition, their emission trend is different as well. Considering that the average CO_2 molecule life is around 100 years, it seems clear that historical emissions are a differentiating factor.

The SCSR concept introduces a way to incorporate geopolitics elements in the discussion, accepting to have into account those facts potentially influential in the air transport development and its impact in the local society, like insularity, economic development, lack of access to the sea, etc.

ICAO launched the process to build up a worldwide CO_2 emissions mitigation system during 2013 Assembly, after having assumed the impossibility of reaching a worldwide agreement around the Emissions Trading System, put on vote, and rejected in the 2007 Assembly. The technical part of the work was taken by the Global MBM

Technical Task Force (GMTF), a new CAEP group, created specifically for this project but, accepting the important political factors involved in the prospective negotiation, an ICAO Council's Environmental Advisory Group (EAG) was set up to review the GMTF provisional results and check the potential acceptability of its proposals in regular meetings from March 2014 up to January 2016.

From the very beginning, it was clear that the offsetting approach was having more supporters than other MBM options and most the GMTF efforts were devoted to find the best way to apply CBDR and SCSR, while producing an efficient way of controlling international aviation CO_2 emissions and help to achieve the ICAO 2020 neutral CO_2 goal. The key elements of discussion were how to classify the different States (volume of international traffic, level of economic development, air transport economic impact), whether to apply the possible payments by airline category or by route features, and the best way of distribute the amount of emissions to be offset among operators, considering their individual growth rates.

A final decision was adopted by the 39th ICAO Assembly on October 6, 2016, with an overwhelming majority. The new MBM, baptized as Carbon Offsetting and Reduction Scheme for International Aviation (CORSIA), was going to be applied to offset the part of international civil aviation CO_2 emissions exceeding the average of 2019–20 levels, starting in 2021. The calculation will be done in net levels, taking into account all other possible offsetting ways, like the voluntary passenger contribution or the use of CDM, included in the Kyoto Protocol, or the certified allowances granted by the use of low-carbon LCA cycle, like biokerosene.

With a very high number of factors to be taken into account, the different solutions should comply with three basic conditions in order to become a feasible system not too difficult to be implemented, monitored, and put into practice: administrative simplicity, environmental integrity, and cost effectiveness.

The solutions for the mentioned key problems make the MBM somewhat complicated, but some intricate tricks were needed in order to ensure the approval by the Assembly:

- The States may decide when to join the system. The first period (2021–26) is voluntary and in January 2018 73 States representing >87.7% of the international RTKs have already declared their intention of participating in CORSIA from the beginning, in the so-called Pilot Period (2021–23). Other 6 (Brazil, Chile, India, Philippines, Russia, and South Africa) will join in the next phase (2024–26) and the others will be added in the Mandatory phase, starting in 2027 and lasting up to 2035. Table 10.4 shows the distribution of the 73 States, corresponding to all the world aviation sectors but South America. These States are trying to get a consensual agreement and are likely to join in the second phase.
- A number of States will be exempted according to four different categories: Least Developed Countries (LDC), Small Island Developing States (SIDS), Landlocked Developing Countries (LLDC), and Low Emitters, not entering in the top 90% RTKs classification and, in any case, with <0.5% of the total RTKs in the year 2018. In Table 10.5 the 2015 State ranking is shown, comparing the volume of international traffic with the total transport developed.
- On distribution, the winning idea was to take into account the routes between States included in the program, independent of the individual airline operating it or its nationality. In such a way, there will be no difference among airlines flying the same route and potential market distortion will be minimized.

Table 10.4 **CORSIA members in January 2018 (*ICAO: CORSIA web page*)**

Region	Number of ICAO states	Number of CORSIA states pilot phase	Member states
European and North Atlantic	55	44	Albania, Armenia, Austria, Azerbaijan, Belgium, Bosnia-Herzegovina, Bulgaria, Croatia, Cyprus, Czech Republic, Denmark, Estonia, Finland, France, FYROM, Georgia, Germany, Greece, Hungary, Iceland, Ireland, Italy, Latvia, Lithuania, Luxembourg, Malta, Moldova, Monaco, Montenegro, Netherlands, Norway, Poland, Portugal, Romania, San Marino, Serbia, Slovakia, Slovenia, Spain, Sweden, Switzerland, Turkey, Ukraine, United Kingdom
Eastern and Southern Africa	24	4	Botswana, Kenya, Namibia, Zambia
North and Central America, Caribbean	21	7	Canada, Costa Rica, El Salvador, Guatemala, Jamaica, Mexico, United States
South America	13	0	
Western and Central Africa	24	3	Burkina Faso, Gabon, Nigeria
Middle East	16	4	Israel, Qatar, Saudi Arabia, UAE
Asia and Pacific	38	11	Australia, China, Indonesia, Japan, Malaysia, Marshall Islands, New Zealand, Papua New Guinea, Singapore, South Korea, Thailand,
TOTAL	191	73	

Table 10.5 **State ranking in international traffic (year 2015) (*ICAO: Annual Report 2015*)**

State	RTK int (million)	Δ 2015/2014	Rank	% of RTK int	RTK total (million)	Rank
United States	61,944	−1	1	10.93	170,585	1
China	54,341		2	9.59	110,185	2
United Arab Emirates	52,501	11	3	9.19	52,101	3
United Kingdom	31,065	3	4	5.48	31,831	4
Germany	30,507	2	5	5.38	31,499	5
Republic of Korea	21,803	4	6	3.85	22,561	8
Singapore	18,616	2	7	3.29	18,616	11
France	18,295	3	8	3.23	22,974	7
Qatar	17,360	21	9	3.06	17,360	13
Netherlands	15,733	−1	10	2.78	15,809	14
Turkey	15,587	13	11	2.73	18,688	10
Japan	15,527	6	12	2.74	23,142	6
Ireland	13,166	14	13	2.32	13,166	17
Canada	13,040	9	14	2.30	18,217	12
Russian Federation	11,635	−1	15	2.05	21,061	9
Australia	9369	5	16	1.65	15,797	15
Spain	9216	15	17	1.63	11,022	19
Thailand	9114	6	18	1.61	10,891	21
Malaysia	8967	−8	19	1.58	10,912	20
India	6994	7	20	1.23	14,344	16
Scandinavia	6925	6	21	1.22	7615	23
Luxembourg	6520	10	22	1.15	6520	26
Switzerland	6102	−8	23	1.08	6119	27
Saudi Arabia	5389	0	24	0.95	6861	25
Philippines	4218	8	25	0.74	5567	28
Brazil	4105	8	26	0.72	12,609	18

Continued

Table 10.5 Continued

State	RTK int (million)	Δ 2015/2014	Rank	% of RTK int	RTK total (million)	Rank
Italy	4030	0	27	0.71	4738	29
Belgium	3596	−1	28	0.63	3596	35
Mexico	3493	20	29	0.62	7209	24
Ethiopia	3454	17	30	0.61	3500	36
New Zealand	3448	4	31	0.61	3885	33
Colombia	3021	8	32	0.53	4206	30
Portugal	2989	−1	33	0.53	3183	37
Chile	2955	1	34	0.52	3993	32
Finland	2955	4	35	0.52	3034	38
Israel	2764	7	36	0.49	2801	39
Indonesia	2683	2	37	0.47	8262	22
Hungary	2664	18	38	0.47	2664	41
Panama	2608	2	39	0.46	2699	40
Austria	2600	0	40	0.46	2613	43
World total (191 States)	566,662	6		100	817,030	

- The offsetting obligation will apply to the airlines flying the routes included in CORSIA. During the voluntary period, the amount to be offset will be proportional to the global performance of the airlines group (100% sectoral) independent of the individual development of each operator. In the 2030–32 period, the individual growth of each airline will account for 20% and the rest will continue being sectoral, while in the last 3 years (2033–35) the individual part will rise to 70%, leaving the other 30% to sectoral.
- The basic reference value will be the average of 2019–20 emissions in the included routes, but States participating in the pilot phase may select for their operators other base year (2021, 2022, or 2023). In the same way, States joining in the first phase may select an emissions base year among 2024–25 or 2026.
- The airlines starting to operate CORSIA routes when the program is already working, may be exempted during the first 3 years, unless its emissions would be 0.1% of the total CORSIA 2020 emissions.
- Special activities like humanitarian, medical, or firefighting flights will be exempted, as will the operations of aircraft with an MTOM lower than 5700 kg, this last caveat created by administrative simplicity reasons.

The approval of CORSIA was received with great general enthusiasm by two reasons. On one side, the almost unanimous consensus put an end to the aviation world community differences on MBMs, clearly demonstrated during the ETS debates. At the same time, it looks like the best tool to complement other efficiency measures based on technical elements, were considered not sufficient to mitigate aviation impact on climate change. Notwithstanding, some voices have made loud statements expressing some doubts on the reach of the system and its potential capability for encouraging airlines to improve their fuel efficiency further than the *business as usual* situation.

The first caveat for CORSIA is the level of coverage. ICAO is competent only in the international field and, taking the 2015 figures as a base (see Table 10.4), this is only 69.35% of the total RTKs. Domestic traffic is particularly important in places like United States, China, Brazil, or Indonesia and will not be included in the regulation. Although it is possible to apply additional measures by national decisions, today the introduction of these policies in the majority of those large domestic aviation markets seems unlikely.

The participation in the system seems quite high, although it is subject to further confirmation. As the time of writing this text, 73 States having declared their intention to join CORSIA represent about 66% of the world RTKs and 87.7% of the international RTKs. From the 40 larger RTK States, only Colombia and Panama, representing about 1.0% of total international RTKs, have not declared their intentions yet, although in the case of the South American countries a possible group decision is being discussed in their civil aviation association CLAC. The limit of 0.5% of the total international RTKs for mandatory integration in the system would reach to Turkey and Colombia, if their participation in 2027 was the same or higher than in 2015. The seven States previously listed will join at the beginning of the mandatory phase (2017).

The way in which international RTKs are defined has been subject of intense discussions. Several parties were defending that each State should be allocated the RTKs corresponding to the flights taking off from its airports, independent of the nationality of the operating airline. However, in the ICAO historical procedures, each State is allocated the RTKs operated by their carriers. In this way, the nationality of the airlines is defining the traffic allocation. This is the system kept in the CORSIA scheme and raises some doubts on the possibility of subsidiary airlines creation in other countries as a way of circumvent the basic regulation.

The system rules cover only the routes between any two States members of CORSIA. Then, with the present levels of traffic, about 40% of the world traffic would be included. The total fuel consumption of world airlines during 2015 was 246 Mton, equivalent to 778 Mton CO_2 emissions. Assuming an average efficiency in the airline domain, 40% of the total would be inside CORSIA after having taking out the domestic sector and the international routes to and from the countries not belonging to the system.

After CORSIA approval, there are still a great number of practical details to be agreed in the next years. The most important of them is to determine the conditions to be complied with by the environmental projects in order to be validated as legal offset sources. After 2005, the introduction of the European ETS has provided some experience in the international carbon markets. During the 2008–16 period (2005–07 was a pilot phase), 1.47 billion offsets were negotiated by European companies. Some other local markets, like California cap-and-trade, Chinese National Carbon Market, or South Korean ETS are also working, but in all of them, the amount of allowed offsetting is limited (in the case of California, up to 8% of the total yearly emissions) and does not envisage a system with unlimited offsetting. ICAO has sent to the member States a State Letter with a draft of the implementation procedures to be approved by June 2018.

An additional and also relevant point is the way to administrate the system, follow the required Monitoring, Reporting, and Verification procedures, collect the payments, and apply them to the environmental programs producing the corresponding offsets. On this point, ICAO is in conversation with the World Bank to share its expertise in this type of multilateral projects.

The cost for the airlines does not appear to be very high. It will depend on the type of offsets to be accepted and their respective prices and, also to which extent it is allowed this offset. At today's present prices the amount would be negligible, but it seems logical to expect emission allowance prices to increase in the future as more permits are demanded to cover offset needs.

Other interesting point to be sorted out is the repercussions of the CORSIA introduction on other MBM already working or in their way to be adopted. It is logical to believe that some countries may decide to align their internal systems with the international standard and they will implement the needed changes in their domestic measures. The most controversial of all existing similar measures is the EU ETS.

After an initial intention of applying ETS to all flights departing or landing in an EU airport, in November 2012 the European Commission decided to limit the system coverage to the flights between two EU airports, leaving intercontinental traffic outside and greatly reducing the amount of emissions to be accounted. The action, nicknamed as "stop the clock," was intended as an interim measure, to give ICAO some additional time for adopting a worldwide MBM and would be revisited after that event.

Once CORSIA has been launched and all the EU States agree on joining it since the initial year of the pilot period (2021), there is a question mark about future EU action. Some stakeholders support the cancelation of ETS to be replaced by CORSIA system, as all the flights presently covered by EU ETS will be included in CORSIA, because the whole EU will participate in that program. However, the EU ETS present limits (5% lower than the average of 2004–06 period emissions) are more demanding than the 2019–20 CORSIA limits. Then, some other voices propose to keep EU ETS or to make a combination of both systems, in spite of the complicated legal and administrative arrangements that the compatibility adaptation would require. Just to give an example, CORSIA system assumes a weight of CO2 emissions of 3.16kg per kilogram of burnt kerosene, while European ETS put the figure of 3.15.

In December 2017, EU published Regulation 2017/2392 maintaining the current ETS until 2023, assuming that the first CORSIA payment will happen in January 2025, covering the excess of emissions in the 2021–23 period. That might mean that European States would pay twice in that period because they are included in both ETS and CORSIA systems.

References

Airbus, 2008. Getting to Grips With A320 Family Performance Retention and Fuel Savings. Issue 2.

Airbus, 2004. Getting to Grips With Fuel Economy. Issue 4.

Air Transport Action Group, November 2010. Beginner's Guide to Aviation Efficiency.

Airport Carbon Accreditation Program (ACAP), www.airportcarbonaccreditation.org.

Airport Council International, 2014–2033. Global Traffic Forecast.

Airbus, 2016–2035. Global Market Forecast.

Alonso, G., Benito, A., Lonza, L., Kousoulidou, M., April 2014. Investigations on the distribution of air transport traffic and CO_2 emissions within the European Union. J. Air Transp. Manag. 36, 85–93.

American Society for Testing and Materials (ASTM), July 2011. Standard Specification for Aviation Turbine Fuel containing Synthesized Hydrocarbons. ASTM D7566.

Aminzadeh, F., Alonso, G., Morales, G., Benito, A., 2016. Analysis of the recent evolution of commercial air traffic CO_2 emissions and fleet utilization in the six largest national markets of the European Union. J. Air Transp. Manag. 55, 9–19.

Ascend database, August 2017. World Airlines Directory. Flight International, pp. 15–21.

ASCENT, 2016. Annual Technical Report. Washington.

Athmann, T., Bjornsson, R., Borrell, P., Thewlis, P., 2008. Geothermal Heating in Airport Runways. Saint Cloud State University, Minnesota.

Avions de Transport Régional (ATR), 2011. Fuel Saving. Toulouse.

Bauen, A., van den Heuvel, E., Maniatis, K., 20 March 2012. Biofuels Flightpath. In: Workshop on Financial Mechanisms for Advanced Biofuel Flagship Plants, Brussels.

Bauen, A., Kyriakos, M., 2012a. Biofuels FlightPath. In: Workshop on Upstream R&D and Innovation for Biofuels in Aviation, Milan. 18 June.

Bauen, A., Kyriakos, M., 2012b. Biofuels FlightPath. In: Workshop on Incentives for Biofuels in Aviation, Brussels, pp. 20. 20 June.

Baxter, G., Sabatini, R., Wild, G., May 2015. In: Sustainable airport energy management: a case study of Copenhagen airport. Proceedings of International Symposium on Sustainable Aviation, Istanbul.

Benito, A., July 2012. In: Sustainable Air Transport: Governance and regulatory framework. Seminar on Strategies for sustainable Development of Air Transport in Brazil. ANAC.

Benito, A., 2009. Technology and maintenance for fuel saving. In: Workshop on Effectiveness and Efficiency: New Aviation Challenges. UIMP, Santander.

Boeing: AERO Magazine. Q4,2007; Q4, 2008; Q2, 2010 n.d.

Boeing, December 2015. B787 Aeroplane Characteristics for Airport Planning D6-58333. Seattle.

Boeing, 2016. Current Market Outlook 2016–2035. Seattle.

Bombardier, 2012. Commercial Aircraft Market Forecast 2012-2031. Toronto.

British Petroleum, June 2017. Statistical Review of World Energy.

CANSO, 2012. ATM Global Environment Efficiency Goals for 2050. https://www.canso.org.

Chester, M., Horvath, A., 2007. Environmental Life-Cycle Assessment of Passenger Transportation: A Detailed Methodology for Energy, Greenhouse Gas, and Criteria Pollutant Inventories of Automobiles, Buses, Light Rail, Heavy Rail and Air. Institute of transportation Studies, University of California, Berkeley.

CLEEN, web page,

Coogan, M.A., 2008. Ground Access to Major Airports by Public Transportation. ACRP4, Transportation Research Board, Washington.

Dickson, N., 2014. Global Emissions Technology. ICAO Environmental.

EASA, 2012. AIR ORS Commercial Air Transport Regulations. Köln.

EIA, 2017. International Energy Outlook.

EIA, 2016. International Energy Outlook.

Embraer, 2015–2034. Market Outlook.

EUROCONTROL, 2015. Long Term Forecast—IFR Flight Movements 2015–2035. Brussels.

European Biofuels Technology Platform, 2014. Biofuels Policy and Legislation. http://bofuelstp.eu/legislation.html.

European Commission, 2005. CONSAVE 2050. Final Technical Report. Brussels.

European Commission, 2016. European Aviation Environmental Report.

European Commission: Flight path 2050. Europe's Vision for Aviation.

European Commission, 9 April 2014. Guidelines on State aid for Environmental Protection and Energy 2014-2020. Brussels.

European Commission, June 2011. Launch of the European Advanced Biofuels Flightpath. Brussels.

European Commission, Single European Sky. https://ec.europa.eu/transport/modes/air/single_european_sky_en.

European Environment Agency, 2013. EU Bioenergy Potential From a Resource-Efficiency Perspective. EEA Report No. 6.

European Union, 2001a. Directive 2001/77/EC on the Promotion of Electricity Produced From Renewable Energy Sources in the Internal Electricity Market (27 September 2001). OJEU. 27 September.

European Union, 2001b. White Paper European Transport Policy for 2010: Time to Decide. COM (2001) 370 final, Brussels, 12 September.

European Union, 17 May 2003. Directive 2003/30/EC on the Promotion of the use of Biofuels or Other Renewable Fuels for Transport (8 May 2003). OJEU.

European Union, 5 June 2009. Directive 2009/28/EC on the Promotion of the Use of Energy From Renewable Sources and Amending and Subsequently Repealing Directives 2001/77/EC and 2003/30/EC (23 April 2009). OJEU.

European Union, Galileo. http://ec.europa.eu/growth/sectors/space/galileo.

Evans, R., Schröder, J., January–February 2017. CORSIA's struggle to offset doubts. In: Flight Airline Business, pp. 48–49.

FORUM-AE, www.forum-ae.eu.

Gnansounou, E., Panichelli, L., Dauriat, A., Villegas, J.D., March 2008. Accounting for Indirect Land-Use Changes in GHG Balances of Biofuels. Working Paper Ref. 437.101, École Polytechnique Fédéral de Lausanne.

Green, J.E., Jupp, J.A., April 2016. CAEP/9-agreed certification requirement for the Aeroplane CO2 Emissions Standard: a comment on ICAO Cir 337. Aeronaut. J., 693–723.

Haering, J., Edward, A., December 1995. Airdata measurement and calibration. In: NASA Technical Memorandum 104316.

Hesse, J.M., July 2015. Climate Change Impact of Airports and Air Navigation Activities. UPM Summer Course, La Granja.

Hind, P., March–April 2014. The Commercial Use of Biofuels in Aviation. Regional International.
Hong Kong Airport Authority, Energy Management. www.hongkongairport.com.
IATA, 2015–2019. Airline Industry Forecast.
IATA, December 2017. Economic Performance of the Airline Industry. www.iata.org/economics.
IATA, 2013. Two Million Tonne per Year Initiative. A Performing Biofuels Supply Chain for EU Aviation. Montreal.
IATA Training, 2017a. Aviation and the Environment. Geneva.
IATA Training, 2017b. Fuel Efficiency and Conservation. Geneva.
ICAO, 2017a. Air Navigation Report 2016. Montreal.
ICAO, July 2005. Annex 2 to Chicago Convention. Montreal.
ICAO, July 2016. Annex 6 to Chicago Convention. Montreal.
ICAO, July 2017. Annex 16 to Chicago Convention. Montreal.
ICAO, 2016a. Annual Report 2015. Montreal.
ICAO, 2010. Annual Report 2009. Montreal.
ICAO, 2013a. CAEP/9 Agreed Certification Requirement for the Aeroplane CO_2 Emissions Standard. Circular 337. Montreal.
ICAO, 2016b. CAEP/10 Proceedings. Montreal.
ICAO, 2014a. CAEP Independent Experts Operational Goals Group. Doc 10021. Montreal.
ICAO, 2013b. Consolidated statement of continuing ICAO policies and practices related to environmental protection. In: A38-18 General Assembly. October.
ICAO, 2014b. EDTO Workshop. Montreal.
ICAO, 2017b. Environmental Report 2016. Montreal.
ICAO, 2015. Global Navigation Plan 2016-2030. Montreal.
ICAO, February 2004. Operational Opportunities to Minimize Fuel Use and Reduce Emissions. Circular 303, Montreal.
ICAO, September 2007. Outlook for Air Transport to the Year 2025.
ICAO, 2008. Performance-Based Navigation Manual (PBN).Doc 9613-AN/937. Montreal.
ICAO, 2011. Review on Sustainable Alternative Fuels for Aviation.
ICCT, February 2016. International Civil Aviation Organization's CO2 Standard for New Aircraft. www.theicct.org.
IEA, 2017. Energy Efficiency.
International Panel on Climate Change (IPCC), 1999. Aviation and the Global Atmosphere. Cambridge University Press.
ITAKA, http://www.itaka-project.eu/default.aspx.
Kharina, A., Rutherford, D., 2015. Fuel Efficiency Trends for New Commercial Jet Aircraft: 1960–2014. ICCT.
Lane, J., 31 December 2013. Biofuels mandates around the world. In: Biofuels Digest. http://www.biofuelsdigest.com.
Lee, J.J., Lukachko, S.P. and Waitz, I.A.: Aircraft and Energy Use, n.d.
Lufthansa Technik, 2017. Maintenance and Overhaul Technologies. https://www.lufthansa-technik.com/.
Mitchell, D., July 2008. A Note on Rising Food Prices. World Bank Policy Research Working Paper no. 4682. Washington, DC.
Mohd Noh, H., Benito, A., Alonso, G., 2016. Study of the current incentive rules and mechanisms to promote biofuel use in the EU and their possible application to the civil aviation sector. Transp. Res. Part D: Transp. Environ. 46, 298–316.
Motherway, B., 2015. The Energy Efficiency Transition. Policies and Perspectives. IEA.
Ninni, A., 2010. Policies to support biofuels in Europe: the changing landscape of instruments. AgBioforum 13 (2), 131–141.

Pilzecker, A., 17 November 2011. Future of Biofuels: An Agricultural Policy Perspective. European Biodiesel Board General Assembly, Brussels.
Rodrigue, J.P., Comtois, C., 2017. The Geography of Transport Systems, fourth ed. Routledge, New York.
Román, J.E., July 2010. Commercial Aviation and the Environment. UPM Summer Course, La Granja.
Rolls-Royce, 2014–2033. Market Outlook.
Schep, E., van Velzen, A., Faber, J., December 2016. A comparison Between CORSIA and the EU ETS for Aviation. CE Delft.
SESAR Joint Undertaking, 2012. AIRE, Summary of Results 2010/2011. Brussels.
Sgouridis, S., Bonnefoy, P.A., Hansman, R.J., 2011. Air Transportation in a carbon constrained world. Long-term dynamics of policies and strategies for mitigating the carbon footprint of commercial aviation. Transp. Res. A 45, 1077–1091.
SWAFEA Final Report, April 2011. Projection of Alternative Fuels (incl. Blending) to 2020+.
Sustainable Aviation Fuel Users Group (SAFUG), August 2010. Governments' Unique Role in Sustainable Aviation Biofuel.
The World Bank, February 2012. Air Transport and Energy Efficiency. Transport Papers.
Union Internationale des Chemins de Fer (UIC), September 2014. Rail Transport and Environment: Facts and Figures. Brussels.
Vecchiet, A., 2011. European Biofuels Blending Obligation. ESSE community. 14 August 2011, http://www.essecommunity.eu.
Vessia, O., Riis Nielsen, K., Grassi, G., Urbancic, N., 16 October 2013. The Future of Biofuels as Alternative Fuel for the Transport Sector. Workshop of the European Parliament Intergroup, Brussels.
Wiesenthal, T., Leduc, G., Christidis, P., Schade, B., Pelkmans, L., Govaerts, L., Georgopoulos, P., 2009. Biofuel support policies in Europe: lessons learnt for the long way ahead. Renew. Sustain. Energy Rev. 13, 789–800.
Yacobucci, B.D., 11 January 2012. Biofuels Incentives: A Summary of Federal Programs. Congressional Research Service R40110, Washington.
Ziolkowska, J., Meyers, W.H., Meyer, S., Binfield, J., 2010. Targets and Mandates: lessons learned from EU and US biofuels policy mechanisms. AgBioforum 13 (4), 398–412.

Index

Note: Page numbers followed by *f* indicate figures, and *t* indicate tables.

D

E